# Advances in Intelligent Systems and Computing

Volume 292

*Series editor*

Janusz Kacprzyk, Polish Academy of Sciences, Warsaw, Poland
e-mail: kacprzyk@ibspan.waw.pl

For further volumes:
http://www.springer.com/series/11156

## About this Series

The series "Advances in Intelligent Systems and Computing" contains publications on theory, applications, and design methods of Intelligent Systems and Intelligent Computing. Virtually all disciplines such as engineering, natural sciences, computer and information science, ICT, economics, business, e-commerce, environment, healthcare, life science are covered. The list of topics spans all the areas of modern intelligent systems and computing.

The publications within "Advances in Intelligent Systems and Computing" are primarily textbooks and proceedings of important conferences, symposia and congresses. They cover significant recent developments in the field, both of a foundational and applicable character. An important characteristic feature of the series is the short publication time and world-wide distribution. This permits a rapid and broad dissemination of research results.

### Advisory Board

Chairman

Nikhil R. Pal, Indian Statistical Institute, Kolkata, India
e-mail: nikhil@isical.ac.in

Members

Rafael Bello, Universidad Central "Marta Abreu" de Las Villas, Santa Clara, Cuba
e-mail: rbellop@uclv.edu.cu

Emilio S. Corchado, University of Salamanca, Salamanca, Spain
e-mail: escorchado@usal.es

Hani Hagras, University of Essex, Colchester, UK
e-mail: hani@essex.ac.uk

László T. Kóczy, Széchenyi István University, Győr, Hungary
e-mail: koczy@sze.hu

Vladik Kreinovich, University of Texas at El Paso, El Paso, USA
e-mail: vladik@utep.edu

Chin-Teng Lin, National Chiao Tung University, Hsinchu, Taiwan
e-mail: ctlin@mail.nctu.edu.tw

Jie Lu, University of Technology, Sydney, Australia
e-mail: Jie.Lu@uts.edu.au

Patricia Melin, Tijuana Institute of Technology, Tijuana, Mexico
e-mail: epmelin@hafsamx.org

Nadia Nedjah, State University of Rio de Janeiro, Rio de Janeiro, Brazil
e-mail: nadia@eng.uerj.br

Ngoc Thanh Nguyen, Wroclaw University of Technology, Wroclaw, Poland
e-mail: Ngoc-Thanh.Nguyen@pwr.edu.pl

Jun Wang, The Chinese University of Hong Kong, Shatin, Hong Kong
e-mail: jwang@mae.cuhk.edu.hk

Tania Di Mascio · Rosella Gennari
Pierpaolo Vittorini · Rosa Vicari
Fernando de la Prieta
Editors

# Methodologies and Intelligent Systems for Technology Enhanced Learning

*Editors*

Tania Di Mascio
Department of Electric and Information
  Engineering
Monteluco di Roio
University of L'Aquila
L'Aquila
Italy

Rosella Gennari
Computer Science Faculty
Free University of Bozen-Bolzanoit
Bolzano
Italy

Pierpaolo Vittorini
Department of Internal Medicine and Public
  Health
University of L'Aquila
L'Aquila
Italy

Rosa Vicari
Federal University of Rio Grande do Sul
UFRGS
Informatics Institute
Porto Alegre
Brazil

Fernando de la Prieta
University of Salamanca
Salamanca
Spain

ISSN 2194-5357          ISSN 2194-5365   (electronic)
ISBN 978-3-319-07697-3   ISBN 978-3-319-07698-0   (eBook)
DOI 10.1007/978-3-319-07698-0
Springer Cham Heidelberg New York Dordrecht London

Library of Congress Control Number: 2014939946

Printed on acid-free paper

Springer is part of Springer Science+Business Media (www.springer.com)

# Preface

Education is the cornerstone of any society and it is the basis of most of the values and characteristics of that society. Knowledge societies offer significant opportunities for new techniques and applications, especially in the fields of education and learning. Technology Enhanced Learning (TEL) provides knowledge processing technologies to support learning activities in knowledge societies. The role of intelligent systems, rooted in Artificial Intelligence (AI), has become increasingly relevant to shape TEL for knowledge societies. However, AI-based technology alone is clearly not sufficient for improving the learning experience. Design methodologies that are based on evidence and involve end users are essential ingredients as well. Stronger with them, AI-based TEL can cater to its learners' tasks and, at the same time, support learning while performing tasks. In brief, we believe that it is the masterful combination of evidence, users and AI that can allow TEL to provide an enhanced learning experience.

The MIS4TEL International Conference is bred and nurtured on the idea of mixing evidence, users and AI for TEL. The conference was built on the experience matured within a series of evidence based TEL workshops: ebTEL 2012, ebTEL 2013, ebuTEL 2013 and iTEL 2013. MIS4TEL expands the topics of the workshops in order to provide an open forum for discussing intelligent systems for TEL and empirical methodologies for their design or evaluation. The conference and these proceedings bring together researchers and developers from industry and the academic world to report on the latest AI-based scientific research and advances as well as empirical methodologies for TEL.

This volume collectively presents all papers that were accepted for ebuTEL 2013 and mis4TEL 2014. All underwent a peer-review selection: each paper was assessed by at least two different reviewers, from an international panel composed of about 30 members of 14 countries. The full program of ebuTEL contains 12 selected papers from 5 countries: Australia, Belgium, Hungary, Italy and United Kingdom. The program of mis4TEL counts 13 contributions from diverse countries such as Australia, Colombia, Croatia, Italy, Mexico, Spain and United States. The quality of papers was on average good, with an acceptance rate of approximately 70%.

The ebTEL and ebuTEL workshops as well as the mis4TEL international conference were launched under the collaborative frame provided by the European TERENCE project (www.terenceproject.eu). The TERENCE project, n. 257410, was funded by the

European Commission through the Seventh Framework Programme for Research and Technological Development, Strategic Objective ICT-2009.4.2, TEL. TERENCE built the first Adaptive Learning System (ALS) with AI technologies and tools for reasoning about stories via educational games.

Last but not least, we would like to thank all the contributing authors, reviewers and sponsors (Telefónica Digital, Indra, INSA - Ingeniería de Software Avanzado S.A (IBM), JCyL, IEEE Systems Man and Cybernetics Society Spain, AEPIA Asociación Española para la Inteligencia Artificial, APPIA Associação Portuguesa Para a Inteligência Artificial, CNRS Centre national de la recherché scientifique and STELLAR), as well as the members of the Program Committee, of the Organising Committee and of the TERENCE consortium for their hard and highly valuable work. The work of all such people crucially contributed to the success of ebuTEL'13 and mis4TEL'14

*The Editors:*

Tania Di Mascio
Rosella Gennari
Pierpaolo Vittorini
Rosa Vicari
Fernando de la Prieta

# Organisation mis4TEL 2014 – http://mis4tel.usal.es

## Steering committee

| | |
|---|---|
| Fernando De la Prieta | University of Salamanca, Spain |
| Tania Di Mascio | University of L'Aquila, Italy |
| Rosella Gennari | Free University of Bozen-Bolzano, Italy |
| Daniel Manrique | Technical University of Madrid, Spain |
| Rosa Vicari | University Fedaral do Rio Grande Do Sul, Brazil |
| Pierpaolo Vittorini | University of L'Aquila, Italy |

## Program Committee

| | |
|---|---|
| Pekka Abrahamsson | Free University of Bozen-Bolzano, Italy |
| Silvana Aciar | Instituto de Informática, Universidad Nacional de San Juan, Argentina |
| Barbara Arfé | University of Padova, Italy |
| Yaxin Bi | University of Ulster, UK |
| Amparo Casado | Pontifial University of Salamanca, Spain |
| Vincenza Cofini | University of L'Aquila, Italy |
| Juan M. Corchado | University of Salamanca, Spain |
| Néstor Darío Duque M. | Universidad Nacional de Colombia, Colombia |
| Juan Francisco De Paz | University of Salamanca, Spain |
| Dina Di Giacomo | University of L'Aquila, Italy |
| Gabriella Dodero | Free University of Bozen-Bolzano, Italy |
| Peter Dolog | Aalborg University, Denmark |
| Richard Duro | Universidade da Coruna, Spain |
| Ana Belén Gil | University of Salamanca, Spain |
| Carlo Giovannella | University of Roma, Tor Vergata, Italy |
| Amparo Jiménez | Pontifical University of Salamanca, Spain |

| | |
|---|---|
| Jon Mikel | Lund University, Sweden |
| Ivana Marenzi | L3S, Leibniz University of Hannover, Germany |
| Marc Marschark | Rochester Institute of Technology, USA |
| António Mendes | University of Coimbra, Portugal |
| Stefano Necozione | University of L'Aquila, Italy |
| Wolfgang Nejdl | L3S, Leibniz University of Hannover, Germany |
| Werner Nutt | Free University of Bozen-Bolzano, Italy |
| Margherita Pasini | University of Verona, Italy |
| Carlos Pereira | ISEC, Portugal |
| Elvira Popescu | University of Craiova, Romania |
| Eliseo Reategui | UFRGS, Brazil |
| Sara Rodríguez | University of Salamanca, Spain |
| Maria Grazia Sindoni | University of Messina, Italy |
| Marcus Specht | Open University of the Netherlands, Netherlands |
| Laura Tarantino | University of L'Aquila, Italy |
| Sara Tonelli | FBK-irst, Italy |
| Fridolin Wild | Open University of the UK, UK |
| Carolina Zato | University of Salamanca, Spain |

## Local Organising Committee

| | |
|---|---|
| Juan M. Corchado | University of Salamanca, Spain |
| Javier Bajo | Technical University of Madrid, Spain |
| Juan F. De Paz | University of Salamanca, Spain |
| Sara Rodríguez | University of Salamanca, Spain |
| Fernando de la Prieta Pintado | University of Salamanca, Spain |
| Davinia Carolina Zato Domínguez | University of Salamanca, Spain |
| Gabriel Villarrubia González | University of Salamanca, Spain |
| Antonio Juan Sánchez Martín | University of Salamanca, Spain |

# Organisation ebuTEL 2013 – http://ebutel.usal.es

## Steering committee

| | |
|---|---|
| Fernando De la Prieta | University of Salamanca, Spain |
| Tania Di Mascio | University of L'Aquila, Italy |
| Rosella Gennari | Free University of Bozen-Bolzano, Italy |
| Pierpaolo Vittorini | University of L'Aquila, Italy |

## Program Committee

| | |
|---|---|
| Pekka Abrahamsson | Free University of Bozen-Bolzano, Italy |
| Mohammad Alrifai | L3S, Leibniz University of Hannover, Germany |
| Barbara Arfé | University of Padova, Italy |
| Anthony Baldry | University of Messina, Italy |
| Vincenza Cofini | University of L'Aquila, Italy |
| Juan M. Corchado | University of Salamanca, Spain |
| Juan Francisco De Paz | University of Salamanca, Spain |
| Dina Di Giacomo | University of L'Aquila, Italy |
| Gabriella Dodero | Free University of Bozen-Bolzano, Italy |
| Peter Dolog | Aalborg University, Denmark |
| Wild Fridolin | Open University of the UK, UK |
| Ana Belén Gil | University of Salamanca, Spain |
| Oscar Gil | University of Salamanca, Spain |
| Carlo Giovannella | University of Roma, Tor Vergata, Italy |
| Kolomiyets Oleksandr | Katholieke Universiteit Leuven, Belgium |
| Ivana Marenzi | L3S, Leibniz University of Hannover, Germany |
| Marc Marschark | Rochester Institute of Technology, USA |
| Stefano Necozione | University of l'Aquila, Italy |
| Wolfgang Nejdl | L3S, Leibniz University of Hannover, Germany |
| Werner Nutt | Free University of Bozen-Bolzano, Italy |
| Margherita Pasini | University of Verona, Italy |
| Elvira Popescu | University of Craiova, Romania |

Sara Rodríguez                    University of Salamanca, Spain
Zsófia Ruttkay                    Moholy-Nagy University of Art and Design,
                                     Hungary
Maria Grazia Sindoni              University of Messina, Italy
Marcus Specht                     Open University of the Netherlands,
                                     Netherlands
Laura Tarantino                   University of l'Aquila, Italy
Sara Tonelli                      FBK-irst, Italy
Carolina Zato                     University of Salamanca, Spain

# Contents

## ebuTEL 2013 - 3rd International Workshop on Evidence Based and User Centered Technology Enhanced Learning

# Study of the Evolution of the Underlying Social Network of Discussion and Generation of New Ideas Applying an Instructional and Learning Model

Fernando Alonso[1], José Font[3], Genoveva López[1], and Daniel Manrique[2]

[1] Departamento de Lenguajes y Sistemas e Ingeniería del Software,
[2] Departamento de Inteligencia Artificial,
Universidad Politécnica de Madrid, Campus de Montegancedo s/n, 28660,
Boadilla del Monte, Madrid, Spain
{falonso,glopez,dmanrique}@fi.upm.es
[3] U-Tad. Centro Universitario de Tecnología y Arte Digital. C/ Playa Liencres, 2,
28290, Las Rozas, Madrid, Spain
Jose.font@live.u-tad.com

**Abstract.** This paper presents empirical results that show how an instructional and learning model might influence the underlying social network of discussion and generation of new ideas and, therefore, knowledge building. This study has been conducted on higher education students taking the third-year program development models course unit, as part of an accredited degree in informatics engineering. A moderately constructivist model with a blended learning approach was implemented in this course unit over the last few years. This combination has improved the academic outcomes achieved by students. In order to analyse what caused this favourable effect, we have analysed the evolution of the underlying social network of generation and discussion of new ideas among students throughout the course unit. We found that some key relationships in this underlying social network change, which suggests that there is a positive impact on knowledge building, learning and, ultimately, student educational achievement.

## 1 Introduction

From the 2005/06 to the 2007/08 academic years, we analysed the academic outcomes of students taking the third-year program development models (PDM) course unit, which is part of an accredited degree in informatics engineering. This course unit was taught according to the traditional face-to-face classroom learning model and assessed by means of a practical assignment and final examination. The results of the analysis conducted over this period suggested that student grades were low. Specifically, the mean grades achieved by students over these three years were 4.28, 4.21 and 4.44 out of 10, respectively. There were no statistically significant differences between the respective grades [1].

As a result of these poor results, we designed a moderate constructivist e-learning instructional model [2] with a blended learning approach (b-learning)

T. Di Mascio et al. (eds.), *Methodologies and Intelligent Systems for Technology Enhanced Learning*, Advances in Intelligent Systems and Computing 292,
DOI: 10.1007/978-3-319-07698-0_1, © Springer International Publishing Switzerland 2014

as advocated by a number of specialists in learning theories [3–5]. This instructional model consists on the eclectic combination of three learning theories —behaviourism [5] (using programmed instruction or computer-assisted learning), cognitivism [6] (using advanced organizers, metaphors, chunking into meaningful parts, and the organization of instructional materials from simple to complex) and constructivism [7] (using a branched design of instruction, rather than a linear format)—, includes psycho-pedagogical prescriptions and combines self-paced learning [8], live e-learning [9] and traditional classroom learning [10].

The outcomes achieved as of then, specifically grades of 5.1 out of 10, suggest that there are statistically significant differences between the grades achieved by students before and after the implementation of our moderate constructivist e-learning instructional model with a blended learning approach. This solution not only reduced underachievement but also increased the number of students feeling that they are well enough prepared to pass [1].

The current aim of our research is to study what causes the beneficial effects of our instructional model with a blended learning approach. As some research results have shown that the social relationships among students play a decisive role in learning environments [11–13], we thought it was important to analyse their evolution throughout the course unit under study.

The leading challenge of a good education is to prepare students for a knowledge building culture. Knowledge building is based on teaching methods focusing on the generation and collective refinement of ideas, involving interactions among people, which is precisely what many of the teaching and learning processes included in our instructional model with a blended learning approach encourage [13, 14]. These interactions, which take place throughout the teaching of the program development models course unit, are a means of coming up with new ideas, which can be refined, assessed and improved through discussion, and stimulate knowledge building [15].

Because of the impact of these interactions on knowledge building, this paper presents empirical results showing that the underlying social network of generation and discussion of new ideas is positively modified (e.g., gets denser) from the start to the end of the course using our instructional model with a blended learning approach. These results could establish a possible association between the network evolution and academic achievement for the course in question.

## 2   The Moderate Constructivist Instructional Model with a Blended Learning Approach

The proposed moderate constructivist instructional model with a blended learning approach structures the educational contents on the basis of the concept of learning objective [16]. A learning objective is the specific knowledge that the learner has to acquire about a concept or skill and the tasks to be performed. It is defined by a set of interrelated learning objects that contain educational contents, a problem to be solved by a workgroup, requiring students to develop cooperative work, and exercises to assess student learning. A collaborative environment is also included, with activities designed to create a social environment that acts as a scaffold for collaborative learning and dialectical constructivism.

This instructional model maps constructivist principles to the instructional design, taking a more pragmatic approach that focuses on the principles of moderate constructivism. It is composed of five phases: analysis, design, implementation, execution and evaluation:

1. The analysis phase defines what to teach. The goal of this phase is to determine the course target knowledge state for the learner in order to properly define the educational contents and the activities that the learner should perform to achieve this state, as well as the emergent technological resources that will support the learning process. Accordingly, the proposed model is a moderate constructivist objective-driven instructional model.
2. The design phase establishes how to teach: the learning process and the educational activities that will take place within this process. It defines the specific educational contents, a problem to be solved by a workgroup that covers the concepts described in the selected educational contents and long-answer questions to assess student learning. For the design of the educational contents, we have used principles based on the content performance matrix and multimedia principles derived from research on information processing psychology within the field of cognitive psychology. These principles further the cognitive processes supported by the memory structures involved in learning.
3. The implementation phase involves building the educational contents, activities and the learning process into a learning management system platform.
4. The execution phase involves learners taking the course. It provides information on the problems encountered and the knowledge acquired.
5. The evaluation phase determines the success of the course and ascertains the learning product quality. For this purpose, information output during execution is gathered and the results are analysed.

Our blended learning process is enacted as a fifteen-week course executed as follows:

1. The course kicks off with a one-day face-to-face session where the learners have the chance to meet each other and the instructor. The instructor presents the learning objectives, discusses the most significant knowledge and tasks to be learned, and describes the interactions there will be through email, chats and forums. Workgroups are also established in this session.
2. Every week there is a two-hour face-to-face session where students ask the instructor questions about the contents they have studied over the last week and discuss problems that they have encountered. The instructor presents the most important contents to be studied over the following week stressing the concepts that are most important or harder to learn.
3. One-hour interactions between learners and between learners and the instructor are informally held every week via chat or forums to consolidate and acquire knowledge.

4. There is permanent email support, where messages are answered within the next day.
5. There is face-to-face support available for the students during six hours per week. Learners can meet the instructor either individually or in groups to clarify contents and receive support on how to overcome difficulties in the problem to be solved by a workgroup.
6. Students have a one-hour group meeting to discuss the educational contents that are being taught during the week and how to apply them to the problem to be solved by a workgroup.
7. An online assessment is held every week throughout a questionnaire.
8. A final face-to-face assessment is held immediately after the course has finished. Learner evaluation takes into account the scores achieved in this examination, weekly online assessments, the solutions given to the problems set to be solved by the workgroup throughout the course and the learner's participation in live e-learning sessions.

## 3   Experimental Study Design

This study was conducted on 81 students of the third-year program development models (PDM) course unit, taught as part of the accredited informatics engineering degree. Subjective measures were gathered using the same two questionnaires that students completed at the start and at the end of the course unit, just before the final examination. Students were asked to rate on a Likert scale from 1 (very low) to 5 (very high) statements on the level of discussion and generation of new ideas with another five fellow PDM students of their choice at both the start and the end of the course unit. The aim is to rate the students' subjective perception of this type of social relationships among peers throughout the course unit taught according to the proposed teaching and learning model. The data collected from both questionnaires was dichotomized, transforming values 1, 2 and 3 to 0; and values 4 and 5 to 1. This transformation is frequent as dichotomous values are frequently used in most common social network analysis programs in order to take into account only frequent and very frequent links [17, 18]. From the data collected and dichotomized in each questionnaire we can observe the evolution of the underlying social network of discussion and generation of new ideas throughout the course unit. To do this, we compared the final dichotomous network with the initial dichotomous network, both built from the dichotomized values gathered from the questionnaires taken at the end and at the start of the course unit, respectively. Each of the nodes of these networks represents a student.

The 81 students that participated in this study were chosen because they have taken the course unit and have completed both questionnaires. The sample was mostly male, the ratio being of about two to three; of the same ethnicity, with very few exceptions, and of a very similar age, about 21 years old. The same two teachers taught all students. Respondents' previous education was not significantly biased as they are all taking the same third-year course unit of the same degree programme. So, an ANCOVA test was not necessary to validate the sample and correct the possible bias in educational level.

## 4    Results

To find out how the proposed instructional model with a blended learning approach might influence the relationship of discussion and generation of new ideas, and therefore knowledge building, we compared the dichotomous networks at the start and the end of the program development models course unit. The results are shown in Table 1, revealing that the final network has considerably more links and is denser than the initial dichotomous network. These two values suggest that this relationship increased significantly as the course unit was taught.

We used social network analysis techniques —social network transitivity coefficient and clustering coefficient— to analyse how the relationship of discussion and generation of new ideas evolved as a result of the course unit teaching/learning processes. The transitivity coefficient is the ratio between the total number of transitive triads and the total number of triads. The clustering coefficient of a network is the average of clustering coefficients of all the network nodes. The clustering coefficient of a node is the ratio between all a node's neighbouring nodes and all possible connections among those neighbouring nodes.

**Table 1.** Percentage evolution of the number of links, density, transitivity and clustering of the final against the initial dichotomous networks for discussion and generation of new ideas

|  | No. of links | Density | Transitivity | Clustering |
|---|---|---|---|---|
| Increment | 18.5% | 16.61% | 28.33% | 43.60% |

The results, shown in Table 1, indicate that all the studied evolution parameters increased substantially, suggesting that there is a clear tendency to form new inter-relationships among students, possibly as a result of the multiple social interactions taking place during the course unit implemented according to the proposed moderate constructivist instructional model with a blended learning approach.

There is also a directly proportional relationship between the increase in new interrelationships among students for the discussion and generation of new ideas, and the knowledge building, leading to better student achievement, as shown in Figure 1. Looking at Figure 1, we find that the density and cohesion index of the group of passed students increased from the start to the end, whereas they de creased for the groups of absent from exam and failed students. The cohesion in dex is the ratio between the existing links among nodes of the same group and links among nodes of different groups. Three groups are defined from the educa tional achievement variable: passed, failed and absent from exam students. A high cohesion index means that the individuals mostly have links with other network members.

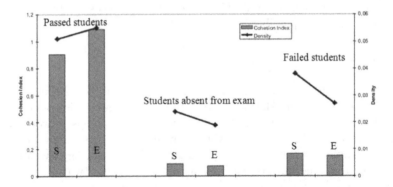

**Fig. 1.** Relationship of discussion and generation of new ideas evolution from the start (S) to the end (E) of the course unit and educational achievement

Interesting results that might corroborate the relationship between the evolution of the social network of discussion and generation of new ideas and academic outcomes of students have been obtained by analysing student academic outcomes against the evolution of the indegree (the number of head endpoints adjacent to a node) of the nodes of the dichotomous networks for this social network from the start to the end of the course unit. The nodes (students) whose indegree decreases or is more or less the same from the start to the end of the course unit belong to students whose average final grade is below 5 out of 10. The nodes whose indegree increases from 1 to 3 passed the course unit. Finally, the average grade of students whose indegree increases by more than 3 links is

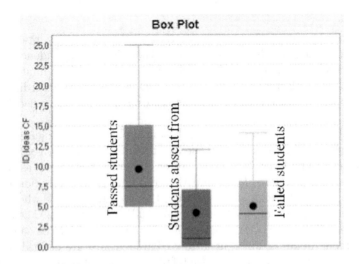

**Fig. 2.** Indegrees of the final dichotomous network for the relationship of discussion and generation of new ideas

nearly 7 out of 10. Figure 2 shows a box plot for the absolute indegrees of the final dichotomous network of discussion and generation of new ideas.

## 5   Conclusions and Future Work

This paper presents an empirical study of the evolution of the underlying social network of discussion and generation of new ideas among students of the third-year program development models course unit, taught as part of an accredited five-year informatics engineering degree. This study was motivated by the belief that applying our moderate constructivist instructional model with a blended learning approach encourages knowledge building and learning through the relationship of discussion and generation of new ideas among students.

It has been shown that the relationship of discussion and generation of new ideas among students was strengthened and the number of links, density, transitivity and clustering increased in the final network (taken at the end of the course unit) with respect to the initial one (taken at the start).

Possible associations between the discussion and generation of new ideas among students and their educational achievement have also been identified. The density and cohesion index of the group of passed students increased notably, whereas they decreased for the groups of absent from exam and failed students. Additionally, the increase in the indegree of students in the underlying dichotomized social network of discussion and generation of new ideas from the start to the end of the course unit is directly proportional to their average academic outcomes (i.e., students with a higher indegree score on average nearly 7 out of 10). On the contrary, students who were less often named as partners with which to discuss new ideas had worse academic outcomes (with an average of below 5 out of 10).

These findings would suggest that learning interactions among students to discuss new ideas during the course unit play a key role. Hence, promoting iterative discussion processes to improve or to advance an idea should be taken into account in instruction.

These results would suggest that there may be a relationship of causality between the application of the proposed instructional model, the positive modification of the social network of discussion and generation of new ideas and the improvement in student educational achievement. This is a potentially interesting conclusion, but requires further research using alternative experimental designs (e.g., with a control group) and statistical analyses. It would also be worthwhile to investigate whether there are other relationships involved. This is proposed as a future line of research.

## References

1. Alonso, F., Manrique, D., Martínez, L., Viñes, J.M.: How blended learning reduces underachievement in higher education: An experience in teaching computer sciences. IEEE Transactions on Education 54(3), 471–478 (2011)

2. Alonso, F., Manrique, D., Viñes, J.M.: A moderate constructivist e-learning instructional model evaluated on computer specialists. Computers & Education 53(1), 57–65 (2009)
3. El-Deghaidy, H., Nouby, A.: Effectiveness of a blended e-learning cooperative approach in an egyptian teacher education programme. Computers & Education 51(3), 988–1006 (2008)
4. Garrison, D.R., Kanuka, H.: Blended learning: Uncovering its transformative potential in higher education. The Internet and Higher Education 7(2), 95–105 (2004)
5. Good, T.L., Brophy, J.E.: Educational psychology: A realistic approach. Longman/Addison Wesley Longman (1990)
6. Anderson, N.H.: A functional theory of cognition. Psychology Press (1996)
7. Jonassen, D.H.: Objectivism versus constructivism: Do we need a new philosophical paradigm? Educational Technology Research and Development 39(3), 5–14 (1991)
8. Ellis, H.J.: An assessment of a self-directed learning approach in a graduate web application design and development course. IEEE Transactions on Education 50(1), 55–60 (2007)
9. Stahl, G.: Building collaborative knowing. In: What We Know About CSCL, pp. 53–85. Springer (2004)
10. Michell, L.: E-learning methods offer a personalized approach. InfoWorld, 174–185 (2001)
11. Cho, H., Gay, G., Davidson, B., Ingraffea, A.: Social networks, communication styles, and learning performance in a cscl community. Computers & Education 49(2), 309–329 (2007)
12. Junco, R., Heiberger, G., Loken, E.: The effect of twitter on college student engagement and grades. Journal of Computer Assisted Learning 27(2), 119–132 (2011)
13. Alonso, F., Manrique, D., Martínez, L., Viñes, J.: Study of the influence of social relationships among students on knowledge building using a moderately constructivist learning model. Journal of Experimental Education (to appear 2014)
14. Perry-Smith, J.E., Shalley, C.E.: The social side of creativity: A static and dynamic social network perspective. Academy of Management Review 28(1), 89–106 (2003)
15. Zhou, J.: Promoting creativity through feedback. Lawrence Erlbaum Associates (2008)
16. Alonso, F., Lopez, G., Manrique, D., Viñes, J.M.: Learning objects, learning objectives and learning design. Innovations in Education and Teaching International 45(4), 389–400 (2008)
17. Huisman, J., Kaiser, F., Vossensteyn, H.: The relations between access, diversity and participation: searching for the weakest link? In: Tight, M. (ed.) Access and Evaluation: International Perspectives in Higher Education Research, pp. 1–28 (2003)
18. De Coster, J., Iselin, A.-M.R., Gallucci, M.: A conceptual and empirical examination of justifications for dichotomization. Psychological Methods 14(4), 349–366 (2009)

# Assisting Students in Writing by Examining How Their Ideas Are Connected

Samuel González López and Aurelio López-López

National Institute of Astrophysics, Optics and Electronics, Tonantzintla, Puebla, Mexico
{sgonzalez,allopez}@inaoep.mx

**Abstract.** Research proposal writing is an arduous process for students and instructors. The proposal must comply with requirements of academic guidelines, and is transformed into a thesis in some cases after several revisions by an adviser. In this paper we present an analyzer to identify the flow of concepts within proposal drafts, with the goal of aiding students to improve their drafting. We propose some novel methods integrated into our analyzer, which were designed considering the transitions of grammar constituents in Problem Statement, Justification and Conclusions sections. We performed experiments on corpora and we could identify anomalous paragraphs. In addition, our results show that graduate students produce better flow of conceptual sequences than undergraduate students.

**Keywords:** Research proposal writing, Draft Evaluation, Entity Grid.

## 1 Introduction

Research proposal writing, sometimes preliminary step of a thesis, is an arduous process for students and instructors. Students have to grasp the key concepts to develop in each section. Instructors have to iterate several times from awkward drafts. In addition, these documents must comply with the features established by institutional guidelines, such as format, structure, legibility, or argumentation. Our efforts aim to assist students in preparing their drafts by examining the flow of their ideas (concepts) in specific sections, using an analyzer. We explore the relationship between paragraphs in Justification, Problem Statement and Conclusion, that is what we call conceptual sequence, that conveys an implicit coherence.

The central contributions of this work are: 1) Developing an analyzer with methods to identify the behaviors in each selected section, based on a local coherence technique previously proposed named Entity Grid (EG); 2) Reporting an analysis of conceptual sequences found in graduate and undergraduate theses; and 3) Showing an implementation to visualize results and provide feedback to students.

We describe a common structure of research proposal drafts (section 3), and we also detail corpora for an experimental study and base technique. Then, we delineate the model of the analyzer (section 4), and develop different schemas to evaluate conceptual sequences, explaining the rationale of the schemas and how they are

T. Di Mascio et al. (eds.), *Methodologies and Intelligent Systems for Technology Enhanced Learning*, Advances in Intelligent Systems and Computing 292,
DOI: 10.1007/978-3-319-07698-0_2, © Springer International Publishing Switzerland 2014

applied in the sections. We also found (section 5) that graduate level texts reach the highest values of conceptual sequences when compared to the undergraduate level after experiments in corpora. Furthermore, we show (section 6) how the analysis is reported to students and feedback is provided.

## 2    Related Work

A document with an appropriate structure presents a clear flow of topics through their paragraphs. For example, in the five paragraph essay paradigm [1], introduction and conclusion share the main topic, this is the theme or subject matter of the essay. The remaining paragraphs in such approach named "body paragraphs" contain details of the essay argumentation and are linked via the main topic. This approach is similar to the proposal drafts, for example in a conclusion section, the paragraphs are connected by the same main topic, and also contains paragraphs that support the results, considering the topic of the problem statement.

The connection between paragraphs involves the interconnection of each of the sentences to the paragraph through its grammar constituents as subject and object. These constituents are observed as a pattern and allow to correctly interpret the information in the text. The sequence presented in the pattern of constituents can characterize specific types of discourses and therefore contribute to the assessment of the quality of the text [2]. In our work, for example, in the Problem Statement section, the set of sentences that integrate each paragraph are interconnected by the same central topic. This flow of connections provides an adequate sense of what the student seeks to address in the research proposal.

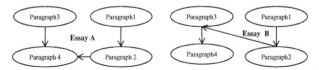

**Fig. 1.** Essay A with a low grade and the Essay B with a high grade of semantic flow

A related study with semantic flow of essays of students learning English is presented in [3]. Authors argue that the paragraphs of the essays have an internal flow, i.e. each paragraph connects with the adjacent paragraph. If paragraphs are connected in ascending order, the essay is suitable from a semantic approach (Fig. 1). They used singular value decomposition, where each paragraph of the essay was represented in a vector space, and then they measure of distance between vectors to determine the semantic proximity of paragraphs, that depended of their topics. They also implemented the Non-negative matrix factorization method, and found that this was suitable for analyzing topic flow. Despite the topic flow in the essays was small, it was present. They concluded that is possible to obtain better semantic flow on collections of essays with more significant quality differences.

In this work, we used a tool based on the Entity Grid (EG) technique to represent discourse, that was proposed to evaluate coherence [4]. The technique generates a representation constructed as a two-dimensional array (EGrid) that captures the distribution of entities in discourse across sentences, where rows correspond to sentences and columns represent the entities of discourse. The cells can have values such as subject (S), object (O), or neither (X). The main idea is that if the object and subject are present in the sentences, the assessed coherence is stronger, assuming that certain types of transition of subject and object are likely to appear in a locally coherent discourse. The EG technique generates a model which is built from a specific corpus and this model is used to evaluate new texts. Similar to our work, we identify the subject and object of sentences with the goal of detecting the flow of concepts, instead of local coherence. In contrast to the original model, our proposal uses text fragments (paragraphs) of the draft to train the model that will assess the rest of fragments of the same document. We evaluated the conceptual sequences of several corpora in Spanish of different graduate and undergraduate levels.

Similar to the previous study, diverse metrics designed to capture various aspects of a well written text were assessed in [5], focusing on text summarization. The linguistic aspects evaluated were grammaticality, non-redundancy, referential clarity, focus, and structure/coherence. One method used was EG, which reached around 90% for accuracy in the grammatical feature. This study was trained on a specific corpus. Instead, we only used fragments of the text to be evaluated to train the model and then evaluate the remaining text with such model.

## 3    Structure of Proposal Draft

A proposal draft consists of a series of elements that have been established by books on methodology and guidelines of institutions. In this work, we selected the sections of Problem Statement, Justification and Conclusions since they are longer texts containing several sentences and this allows to generate paragraph models. In contrast, an Objective or Hypothesis would not generate a model, since they are commonly composed of one sentence.

We gathered several corpora of the different elements, distinguishing in these corpuses two kinds of student texts: graduate proposal documents, and undergraduate documents, such as theses. The corpora consist of 240 collected samples, 120 samples for graduate (G) and 120 for undergraduate (U) level, with forty samples of each of the sections: problem statement, justification, and conclusion. The first kind of texts included documents of master and doctoral level. The second kind included documents of Bachelor and Advanced College-level Technician degree. The corpus focuses on the domain of Computing. Each item of the collected corpora is a document that had been evaluated at some point by a reviewing committee.

## 4    Model of Analyzer

Our model incorporates different schemes (methods) to evaluate the flow of conceptual sequences and each is applied depending on the section that is being assessed. For Problem Statement and Conclusions, First Paragraph of Reference and

Evaluation of Nearest Neighbor (ENN) methods are adequate. For the Justification section, Cascade Evaluation, ENN, and Auto-evaluation of Paragraphs are pertinent (Fig. 2).

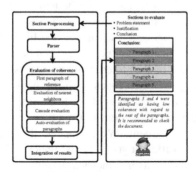

**Fig. 2.** Analyzer model

Section preprocessing is done in two main tasks. The first focuses on segmenting each section (Justification, Problem Statement and Conclusions) into paragraphs, i.e. sequences of sentences bound by line feeds. EG tool requires as input, the text in a Treebank format. The second task is a translation from Spanish to English[1], since our corpora is in Spanish. The result of the translation enables to process the text with an English parser, in particular we are using Stanford[2] (Currently, the parsers for Spanish do not adhere to the Treebank tags).

The method for analysis emerged after analyzing the behavior of the transitions on the EGrid of 10% of samples of the three sections of the graduate corpus. The analysis was done employing the Coherence Toolkit[3], using basically the commands Train and DiscriminateRand (DR)[6]. The first command generates the models, and the second uses the generated model and evaluates the paragraphs. The second command performs a binary discrimination task, which tests the ability of the generated model to differentiate between a document in its original order and a random permutation of that document, and produces results in terms of Accuracy and F-measure. So, we generate the model of a paragraph and then applied DR to evaluate a paragraph of the same document with that model. The idea was that the model could predict whether the paragraph was related to the model: the higher the F-measure and accuracy, the stronger the relationship, i.e. there is evidence that the two paragraphs have a flow of conceptual sequence. Otherwise, the evaluated paragraph is not connected. We now describe each of the schemes.

**First Paragraph of Reference (FPR)** When analyzing Conclusions, we observed that most of the transitions appear as first paragraph entities, i.e. entities identified by the tool in the first paragraph are further shown in the rest. Fig. 3 depicts that the entities identified in the first paragraph are within the dotted circle, the rest corresponding to the second paragraph. In addition, the remaining paragraphs also

---

[1] We used Google Translator.
[2] http://nlp.stanford.edu:8080/parser/
[3] http://cs.brown.edu/~melsner/manual.html

included same entities as the first. Note that the transitions appeared in a sequence (S, O, X, -). For instance, in the marked square zone, the entity "systems" was identified in the second sentence as subject. Later, in the third and fourth sentences appeared as the object, and finally was identified as object in the eleventh sentence. These transitions provide evidence that most paragraphs are adequately connected in term of concepts.

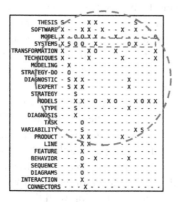

**Fig. 3.** EGrid of a Conclusion

The FPR scheme begins by generating a model of the first paragraph. Then, subsequent paragraphs are evaluated with such model, expecting that they get a positive value. The results provided by the tool are in the range from 0 to 1, where zero indicates a random flow of conceptual sequence, in some cases a null flow. A result of one implies the existence of a relationship between the model and the evaluated text, in this case a subsequent paragraph, i.e. the flow of conceptual sequences is strong. After comparing each result obtained by the EGrid of the graduate corpus with the content of each Conclusions section, we found a strong flow to those paragraphs that showed a value higher than 0.5 and a weak flow to those below that value. In assessing the subsequent paragraphs, we expect the results to be above zero, which would show that the section have a fit conceptual sequence.

We illustrate how a conclusion is analyzed (Table 1). First, we assessed the conclusions with the FPR method, generating the first model of paragraph 1, which was used to evaluate the remaining paragraphs.

**Table 1.** Conclusion section of undergrad corpus

| Paragraphs | | | | | | | |
|---|---|---|---|---|---|---|---|
| 1 | 2 | | 3 | | 4 | | Method |
| | ACC | FM | ACC | FM | ACC | FM | |
| Paragraph of Model 1 | 0 | 0 | 0.5 | 0.66 | 0 | 0 | FPR |
| | 0.5 | 0.66 | Model 2 | | 0.9 | 0.9 | ENN |

ACC=Accuracy and FM = F-measure

As a result, it was found that paragraph 2 and 4 showed a null connection to the first one (referred hereon as null paragraphs). Instead, paragraph 3 displayed a good connection to the first, getting a value of 0.5 in accuracy and 0.66 in F-measure. From these preliminary results, we assumed that the conclusion was partially disconnected.

Afterwards, we sought to relate the null paragraphs with its prior and subsequent neighbors with the ENN method (detailed below). In this case we only had the option to build the model of paragraph 3. As a result of evaluating paragraph 2 with paragraph 3 (model 2), we obtained a value of 0.66 in F-measure, which indicates that there is a relationship between them. Later, paragraph 4 was evaluated with model 2, finding an F-measure of 0.92 that indicates a strong connection between paragraph 3 and 4. Light gray shading indicates that a close left neighbor has been found. From these results, the evaluated section shows evidence of a fit conceptual sequence. In the case where a close neighbor could not be found, we would infer that the conclusion is not properly connected and has to be restructured.

**Evaluation of Nearest Neighbors (ENN)** is a method designed to evaluate null paragraphs identified after being examined with the FPR scheme. So, its main purpose is to relate null paragraphs with their prior or subsequent neighbor.

The first step is to evaluate the null paragraph with the prior neighbor paragraph model. If there is no relationship, we proceed to evaluate the paragraph with the subsequent paragraph model. Finding a relationship between the null paragraph and its neighbor paragraphs, we detect a connection with the rest of the conclusions. Also, all paragraphs of the evaluated section would show some flow of concepts (see Fig. 4).

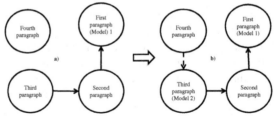

**Fig. 4.** a) FPR results; b) ENN results

Fig. 4a shows four paragraphs of a Conclusions section evaluated by the FPR method, with the first three paragraphs having an acceptable flow (strong connecting lines).The fourth paragraph is null and shows no connection. When applying the ENN method on the null paragraph, we obtained a connection to its close prior neighbor (dotted line, Fig. 4b), enabling the interconnection of the entire section and the null paragraph is actually connected.

Table 1 also illustrate the application of ENN method. Notice that it is not necessary to use the ENN scheme in paragraphs where the results of the evaluation were higher than zero, since they are already showing a relationship with the first.

**Cascade Evaluation** (CE) scheme evolved from what was observed in the EGrid regarding the Justification section. EGrid shows that transitions are distributed and do not concentrate on one position. A comparative review between the EGrid and the evaluated original text of the Justification allowed to find that some paragraphs presented a sequentially thematic relationship.

For instance, Fig. 5 shows a partial EGrid of the third and fourth paragraphs of a Justification of a grad text. The third paragraph contained the entities "challenge" and "objective", which are identified as subject and object, respectively. Later, the same entities appear with subject roles in the fourth paragraph. This similarity in the entities revealed that the paragraphs are indeed directly related.

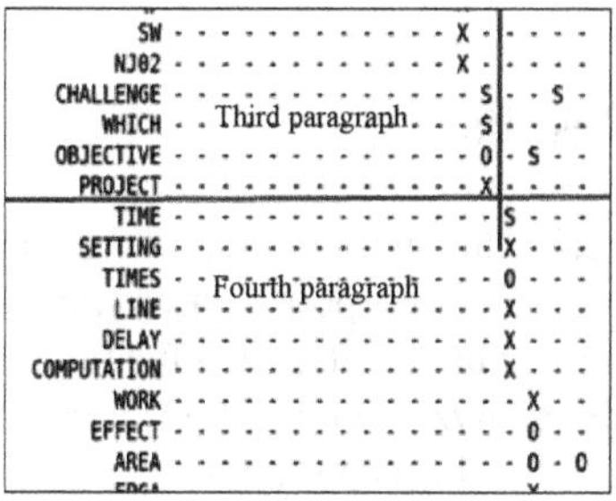

```
SW          · · · · · · · · · · · X ·|· · · ·
NJ02        · · · · · · · · · · · X ·|· · · ·
CHALLENGE   · · · · · · · · · · · · S|· · S ·
WHICH       · · Third paragraph. · · S|· · · ·
OBJECTIVE   · · · · · · · · · · · · O|· S · ·
PROJECT     · · · · · · · · · · · X|· · · ·
TIME        · · · · · · · · · · · · ·|S · · ·
SETTING     · · · · · · · · · · · · ·|X · · ·
TIMES       · · Fourth paragraph · · · O · · ·
LINE        · · · · · · · · · · · · X · · ·
DELAY       · · · · · · · · · · · · X · · ·
COMPUTATION · · · · · · · · · · · · X · · ·
WORK        · · · · · · · · · · · · X · ·
EFFECT      · · · · · · · · · · · · O · ·
AREA        · · · · · · · · · · · · O · O
```

**Fig. 5.** EGrid of a Justification

This behavior looks like a thematic window between two paragraphs, showing a relationship to the preceding paragraph. This window moves between the rest of paragraphs. The CE method works by generating the model of the first paragraph and this model is used to evaluate the second paragraph. Subsequently, we generate a model of the second paragraph, and the third paragraph is evaluated with this model. This process is repeated for all paragraphs of the Justification.

An example of the undergraduate corpus evaluation of the Justification section is shown in Table 2 that shows that the models appear as a stairway. Paragraph 2 shows a value of 0.55 for F-measure when evaluated with the model of the first paragraph, and paragraph 5 has a relationship to paragraph 4 with a value of 0.66.

**Table 2.** Justification section of undergrad corpus

| Paragraphs | | | | | | | | | |
|---|---|---|---|---|---|---|---|---|---|
| 1 | 2 | | 3 | | 4 | | 5 | | Method |
| Model 1 | ACC | FM | ACC | FM | ACC | FM | ACC | FM | |
| | 0.55 | 0.55 | | | | | | | |
| | Model 2 | | 0 | 0 | | | | | CE |
| | | | Model 3 | | 0.9 | 0.92 | | | |
| | | | not found a close neighbor | | Model 4 | | 0.5 | 0.66 | ENN |
| | | | Paragraph with 2 sentences | | | | | | AP |

ACC=Accuracy and FM = F-measure

Afterwards we applied the ENN method, but did not find any neighbor paragraph. Finally, the AP method (below) was not applied because the number of sentences was not enough. This result allowed to identify that in the middle of the Justification there was a null paragraph, i.e. disconnected from the remaining text.

**Auto-evaluation of paragraphs (AP)** was designed to evaluate null paragraphs that remained after being evaluated with any of the previous methods. We decided to assess the paragraphs individually, hoping that they were at least connected properly within, even though they were not related to the other paragraphs. The first step is to divide each of the paragraphs in two parts, but only those paragraphs with at least four sentences, since the tool does not generate models for one sentence. Afterwards, we apply the FPR method, i.e. generate the model of the first "paragraph" and then the second is evaluated with this model. So, we obtain a value of connection for the individual paragraph.

## 5     Experimental Results

The objective of the experiments was to apply the analyzer and the previously designed evaluation schemes on our corpora. In this way, we generated a diagnosis of both levels in the Problem Statements, Justifications and Conclusions. In the experiments, we applied the analyzer with the different methods included. We used the remaining 90% of our corpus, in each section.

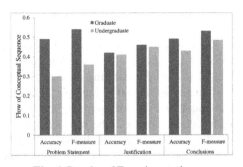

**Fig. 6.** Results of Experimentation

Fig. 6 summarizes the results in terms of average Accuracy and F-Measure values provided by the tool, for the three sections. These results show that grad students' paragraphs from our corpus are better linked than those of the undergrad students, being wider the difference for Problem Statements and closer for Justification. The results were qualitatively validated in the original text, i.e. we looked for the relation between paragraphs as found by the methods, and they coincided.

## 6     Analysis Visualization and Feedback

We want that students perceive visually the different levels of relationship between the paragraphs in their texts. This will allow them to have a general idea of the text regarding the feature of conceptual sequence.

Fig. 7 shows the web interface of the conceptual sequence analyzer, that was designed to enable the student to evaluate his draft in the three sections. We use different shades of blue to show the flow of concepts in the document, where a dark blue indicates higher flow in paragraph, while a light blue represents a low flow. Moreover, when a paragraph is not colored, it means that is disconnected from the rest. The analyzer provides instructions to the student on how to interpret the different shades of color, and shows a report of the analysis of each examined paragraph, interpreting the reached levels of concept flow. Also, the analyzer sends feedback to the student with hints to improve the text, if necessary (Fig. 8).

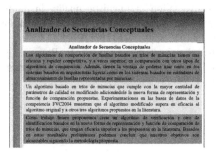
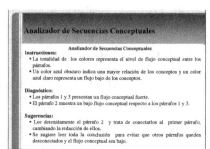

Fig. 7. Analyzer (in Spanish)                   Fig. 8. Feedback (in Spanish)

# 7    Conclusions

Assessing the flow of ideas in drafts is a complex task for computers, and sometimes even for humans. A contribution of our research is that the Entity Grid tool allows to capture the grammatical essence of main concepts without details. In our work, we presented schemes to assess conceptual sequences of drafts.

With the EG tool, we started from its basic operation and built methods that capture the behavior of the different evaluated sections. These methods were incorporated into an analyzer, which was helpful to make examine the flow of conceptual sequences on corpora. We identified that the graduate students from our corpora outperformed the undergraduate students in their writings.

We are currently exploring language models for characterizing other sections, e.g. Objectives, Research Questions, or Hypothesis. We foresee an experiment that will include a pilot test with a control and experimental group of students. This test can provide an insight on whether there are improvements in students in the process of drafting when supported by the analyzer.

**Acknowledgments.** This research was supported by CONACYT, México, through the scholarship 1124002 for the first author. The second author was partially supported by SNI, México.

# References

1. Davis, J., Liss, R.: Effective Academic Writing 3, The essay. Oxford University Press (2006)
2. Webber, B., Egg, M., Kordoni, V.: Discourse structure and language technology. Nat. Lang. Eng. 18, 437–490 (2011)

3. O'Rourke, S., Calvo, R.: Analysing Semantic Flow in Academic Writing. In: Proceedings of the 2009 Conference on Artificial Intelligence in Education: Building Learning Systems that Care, pp. 173–180. IOS Press, Amsterdam (2009)
4. Barzilay, R., Lapata, M.: Modeling local coherence: An entity-based approach. Comput. Linguist. 34, 1–34 (2008)
5. Pitler, E., Louis, A., Nenkova, A.: Automatic evaluation of linguistic quality in multi-document summarization. In: Proceedings of the 48th Annual Meeting of the Association for Comput. Linguist (ACL 2010), Stroudsburg, PA, USA, pp. 544–554 (2010)
6. Elsner, M., Charniak, M.: Disentangling chat with local coherence models. In: Proceedings of the 49th Annual Meeting of the Association for Comput. Linguist.: Human Language Technologies, Stroudsburg, PA, USA, vol. 1, pp. 1179–1189 (2011)

# Extreme Apprenticeship Meets Playful Design at Operating Systems Labs: A Case Study

Vincenzo Del Fatto, Gabriella Dodero, Rosella Gennari, and Naomi Mastachi

Free University of Bozen-Bolzano, Piazza Domenicani 3, 39100 Bolzano, Italy
{vincenzo.delfatto,gabriella.dodero,rosella.gennari,
naomi.mastachi}@unibz.it

**Abstract.** The extreme apprenticeship instructional methodology, recently born in Scandinavia, serves to organise education in formal contexts, such as university courses. The fundamental idea is that a new task is learned by apprentices by looking at the master who is performing it, and then repeating the task under his or her guidance. Continuous feedback and learning by doing are key principles of extreme apprenticeship. However, in e-learning contexts, the direct contact with the master may be missing. Then engagement of students with learning material becomes a challenging goal to achieve when designing the material. In this paper, we see how extreme apprenticeship and playful design were combined for designing the learning material of the laboratories of a 'boring' university course, namely, operating systems. A preliminary analytic evaluation concludes the paper showing the viability of the blended approach.

**Keywords:** Technology Enhanced Learning, Methodologies for Gamified or Game-based Learning, eXtreme Apprenticeship, Playful Interface Design, Interface Design for Learning.

## 1 Introduction

The fundamental idea of the *eXtreme Apprenticeship* (XA) instructional methodology is that apprentices learn by looking at the master performing tasks, and trying them over and over, in small chunks, under the master constant guidance. XA has been successfully applied in science curricula and, more recently, in computer science curricula in Finland, where it was born, and elsewhere. However, in e-learning contexts, XA instructors are not necessarily present. Then instructors face a challenging problem: how to engage learners with learning material, if they cannot watch instructors practicing it. *Playful design* (PD), borrowing elements from game design, can help XA instructors in designing engaging material for e-learning contexts. This paper purports the idea of blending XA and PD in such a context. It does so by showing how the learning material of the labs of a "boring" course was organised using XA principles, and how the interface of the material, made of videos, was designed with PD principles in mind. The paper starts by providing the necessary background concepts. It moves on presenting how XA and PD were applied for designing the lab material and its interface.

T. Di Mascio et al. (eds.), *Methodologies and Intelligent Systems for Technology Enhanced Learning*, Advances in Intelligent Systems and Computing 292,
DOI: 10.1007/978-3-319-07698-0_3, © Springer International Publishing Switzerland 2014

The paper ends by showing results from a preliminary evaluation of the interface with inspection methods, providing novel ideas for future work.

## 2   Background

This section provides the reader with background information concerning XA, for organising learning material, and PD, for designing their interface.

### 2.1   Extreme Apprenticeship

Recently a new approach to teaching introductory programming courses has received much attention, namely XA. It has originally been developed at the University of Helsinki, from where it has started to spread around [1,2,3]. XA is a comprehensive approach for organising education in formal contexts, and it is based on Cognitive Apprenticeship (CA)[4]. In CA, a new task is learned by apprentices, looking at the master who is performing it, and then repeating the task under his or her guidance. So far, XA has been applied to teaching Mathematic topics such as Linear Algebra and Logic [5], as well as Computer Science subjects such as Introduction to Programming, Algorithms, and Operating Systems [6]. Basic principles of CA are:

1. *learning by doing:* the craft can only be mastered by actually practicing it, as long as it is necessary. So, students must do many exercises, which have been designed to simultaneously build both skills and knowledge;
2. *formative assessment via bidirectional feedback:* the learning process is effective by means of continuous, bidirectional feedback. Teachers must be aware of successes and challenges of learners, giving them, as frequently as possible, even small signals of encouragement.

Results achieved so far by adopting XA are impressive, reducing drop-out rate, increasing exam grades, and achieving high retention of learned skills. Such achievements rely upon flexible arrangement, in the spirit of Extreme Programming, of tutoring on-demand. Guidance to students in XA is based on Vygotsky's idea of scaffolding [7]: students are given just enough hints to proceed, boosting in this way their ability to solve the proposed task. Scaffolding progressively fades over time, as the students begin mastering themselves the task.

Much emphasis is given by CA (and XA) on the role of exercises. They are conceived for "teaching the same material (as lectures) but in an exploratory fashion" [8]. This exploratory approach fosters intrinsic student motivation, which in turn improves student performance. XA is aware that difficulties in an assignment may result in killing the motivation of the average-to-weak students, resulting in them dropping out. By providing students with many weekly exercises, each of them requiring to master a minimum amount of new material on top of previous exercises, students acquire new skills by confronting themselves with a measurable amount of work to be done.

Another crucial factor to students achievement is the level of comfort, which is based on self-esteem and self-efficiency [9]. Students in XA-based courses assess their own self-efficiency by looking at the amount of daily work performed, in terms of number of solved exercises. Scaffolding contributes mostly to self-esteem, where expert's feedback always provides some means to improve students' perception of self. As an example of the latter, a positive feedback must contain a sufficient grade, but quite often there is also some word of encouragement ("Well done!") or just a smiley ("☺").

### 2.2 Playful Design for Engaging Learners

When it comes to preparing learning material to be consumed online, e.g., on e-learning platforms, learner experience design comes into play. This requires, first of all, to treat learners as users of learning products, and hence to design for their *User eXperience* (UX). A modern trend in UX design sees products designed like games, that is, "gamified" [10]. In its most common acceptation, *gamification* means properly using game-based elements for a non-game product and in a non-game context in order to engage people in the product itself.

Despite its variety of meanings and applications, gamification in UX design always requires playful engagement as key goal [11,12,13]: as Kumar and Herger point out in [14], "while effectiveness, efficiency, and satisfaction are worthy [usability] goals, gaming and gamification extend and add increased engagement to these goals". Products get structured into missions, which in turn contain challenges and other game elements, such as rewards, that get designed according to the target types of users, conceived as players, and their motivations to play. When products are intended for learning, challenges and award-winning competitions are often added to make them less "boring rote".

However, as claimed in Ch. 5 in [15], "in each and every case, the interface design has played some part in the success or failure of the experience". For designing interfaces, one has at disposals a number of guidelines such as Nielsen heuristics [16] for promoting usability, in the sense of effectiveness, efficiency and satisfaction.

Moreover, as argued in [15], specific game mechanics and aesthetics principles for engaging learners should be added to the interface designers' toolkit: game-mechanics principles help designers to employ game elements such as rewards and progressive challenges; aesthetics principles help designers to support "clarity, communication, comprehension, and emotion" and the better the aesthetics of an interface is "the more credible users will believe the content to be".

## 3 XA for Operating Systems

All the learning material described in this section originates from the *Operating Systems* (OS) course at the Free University of Bozen-Bolzano. The course was offered in the Bachelor programme with the XA methodology applied to the Operating Systems labs for three academic years, from 2011 till 2013.

During those three years, the OS lab and material were organised following XA principles. Tutoring was available in labs overall 6 hours per week. Bash scripting, covering the contents of [17], was split into many small exercises, organised into thematic weekly units, which required learners to progressively acquire specific skills. Bash exercises were distributed in plain text format.

During lab activities, however, students experienced occasional lack of concentration in reading such exercise texts, that resulted in trivial mistakes, especially when lab hours were scheduled in late afternoon. Several students could not regularly attend labs, e.g., due to overlaps with other courses. To avoid such issues, many students asked to solve exercises at home, at their most convenient times, and in a more quiet environment. They asynchronously uploaded their final solutions to the university e-learning portal, where teachers periodically downloaded and assessed them. This implied to deliver the lab in a blended fashion, scaffolding students with asynchronous feedback.

However, working alone at home may cause a lack of motivation, mainly due to the loss of synchronous interaction with peers and teachers. Our main goal thus became how to keep students engaged when working at home. This lead to the following high-level requirement: *creating more engaging self-study material*.

Moreover, the material on the e-learning portal reached a larger population of university students, including full-time workers. The second main high-level requirement for the material was then: *making online material that is usable and accessible by diverse types of university students, different in terms of background knowledge and attention span*.

For this reason, in Fall 2013 we planned to redesign the learning material, turning text exercises into videos, to be made available on the university e-learning portal, where tutors are not always available. The remainder of the paper describes how the videos were designed so as to motivate diverse university students to consume them.

## 4  Playful Design for Operating Systems

Videos allow us to watch things happening in motion and can arouse emotions more easily than static material. As stated in [15], "studies have shown that watching, and even just thinking about, physical things activates the same parts of the brain that are activated when we actually do those things". Videos were thus chosen for showing how the XA instructor performs with the learning material and to train learners as apprentices to that. In this section, we focus on how we designed such videos following specific design principles. In the end, we briefly sketch a preliminary analytic evaluation of the videos and its main results.

### 4.1  Design Choices

In designing videos for XA learning, we considered traditional usability principles such as Nielsen heuristics [16]. For instance, help for less experienced learners is provided via tips and notes as shown in Fig. 2. Navigation through the learning

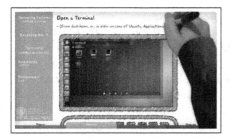

**Fig. 1.** Video screen-shots with tasks for learners in English (left) and German (right)

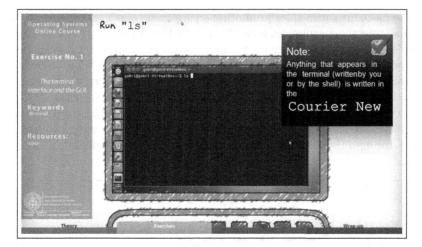

**Fig. 2.** A video screen-shot showing a note for less experienced learners and the usage of animation for showing learners how to run "ls"

material is designed so as to let learners experience control and freedom in moving through the material. As Fig. 3 shows, navigation tabs at the bottom of the interface show learners what types of material the video is currently showing them, e.g., theory or exercise. The left-side of the interface, on the other hand, show learners at what point they are in their learning path by listing: the title and identifier of the video, its key concepts, its prerequisites (resources). Moreover, accessibility was a key concern and multi-modality is implemented so as to reach as many learners as possible, matching their world and metaphors. For instance, videos are both spoken and written in English, German and Italian, the three official languages of the Free University of Bozen-Bolzano. See Fig. 1. Users with reading problems can run the video in slow motion.

We concentrated in particular on aesthetics, which is deemed relevant for engaging students in learning with a positive attitude and making them perceive the learning source more creditable. As claimed by Norman [18], finding something attractive brings "a more positive mood", to the point that students become more willing to tackle problems. Aesthetic design for learning counts specific strategies [15]:

1. reducing overload, that is, striving for minimalism;
2. guiding attention to the relevant part of the interface, e.g., by positioning objects most relevant to learning in the centre of the interface as shown in all figures;
3. supporting visual perception, e.g., avoiding colour and texture faux pas;
4. implementing visual representations that promote visual learning, in particular: (1) visual cues for understanding and remembering, e.g., see Fig. 1; (2) still representations for concepts, such as the representation of the terminal as in Fig. 3; (3) animation for showing relevant tasks, such as the animation that shows how to run the "ls" command in Fig. 2, but sparingly.

Game-mechanics principles promoting engagement were also considered, such as segmenting the learning material into topic-based progressive tasks, turned into videos that last no longer than 5 minutes, so as to maintain concentration. This and the left-side navigation bar allow learners to choose topics relevant to them, and to review more easily the material, enabling shortcuts to learning. Feedback concerning exercise processing, meant for the 'entire class' [19], and discovery elements are also used to keep learners engaged, as illustrated in Fig. 4. Moreover, lab assistants are displayed as talking avatars for guiding and assisting learners through the learning material, as shown in Figg. 3 and 4.

**Fig. 3.** A video screen-shot showing the avatar for attracting the attention towards the video learning goal, and the visual representation of a terminal, fostering recognition rather than recall

## 4.2   Preliminary Evaluation Results

For evaluating the video interface, we did a preliminary expert review. Reviewers were: two experts of usability and game design; two of special needs education.

**Fig. 4.** A discovery element for engaging

Results are rather uniform across reviewers. All reviewers found videos accessible and promoting multiple means of representation. Animations were considered relevant and not distracting from the main task.

Usability experts found room for improvement. Both noticed that visual cues were not always consistently used. For instance, the writing hand of Fig. 1 is mainly used in relation to tasks that learners should perform. At points, albeit rarely, the hand is used for other purposes. The avatar has a clear role when guiding through the learning material. That said, at points the avatar is used where the writing hand should be used to assign tasks. Moreover, juicy feedback was also perceived as crucial. According to one usability reviewer, however, learners should be more challenged to tackle exercises before watching the feedback for their resolution in the wrap-up section: adding an activity that requires a learner to stop, run an exercise and input its resolution between segments of passive information should help in keeping learners even more engaged.

## 5   Conclusions

This paper advocates the use of PD for designing learning material that is organised with XA principles. The paper explains the adopted XA and PD principles for engaging learners in e-learning contexts, where the XA instructor may not be present and engagement becomes critical. It does so by illustrating such principles in the context of a case study, namely, that of an operating systems lab. Lessons learnt from the study shows the viability of the approach, and pave the way for future work at the intersection of XA and PD in e-learning contexts. On the other hand, the availability of self-study engaging material allow to support other types of students such as high school ones, who would like to learn about courses offered by the local university before enrolment, and lifelong learners. In particular, we are designing contextual inquiries with high school students and teachers for testing the use of OS videos in in their specific context.

# References

1. Vihavainen, A., Paksula, M., Luukkainen, M.: Extreme apprenticeship method in teaching programming for beginners. In: Proc. of the 42Nd ACM Technical Symposium on Computer Science Education, SIGCSE 2011, pp. 93–98. ACM, New York (2011)
2. Kurhila, J., Vihavainen, A.: Management, structures and tools to scale up personal advising in large programming courses. In: Proceedings of the 2011 Conference on Information Technology Education, SIGITE 20111, pp. 3–8. ACM, New York (2011)
3. Vihavainen, A., Paksula, M., Luukkainen, M., Kurhila, J.: Extreme apprenticeship method: key practices and upward scalability. In: Proceedings of the 16th Annual Joint Conference on Innovation and Technology in Computer Science Education, ITiCSE 2011, pp. 273–277. ACM, New York (2011)
4. Collins, A.: Cognitive apprenticeship. In: The Cambridge handbook of the learning sciences, pp. 47–60. Cambridge University Press (2006)
5. Hautala, T., Romu, T., Rämö, J., Vikberg, T.J.: Extreme Apprenticeship Method in Teaching University-Level Mathematics. In: Proc. of the 12th International Congress on Mathematical Education, ICME 2012 (2012)
6. Dodero, G., Cerbo, F.D.: Extreme apprenticeship goes blended: An experience. In: Giovannella, C., Sampson, D.G., Aedo, I. (eds.) ICALT, pp. 324–326. IEEE (2012)
7. Vygotskiĭ, L.: Mind in society: The development of higher psychological processes. Harvard Univ Pr (1978)
8. Roumani, H.: Design guidelines for the lab component of objects-first CS1. In: Proceedings of the 33rd SIGCSE Technical Symposium on Computer Science Education, SIGCSE 2002, pp. 222–226. ACM, New York (2002)
9. Bergin, S., Reilly, R.: The Influence of Motivation and Comfort-level on Learning to Program. In: Proc. of the 17th Workshop on Psychology of Programming, PPIG 2005 (2005)
10. Ferrara, J.: Playful Design. Rosenfeld Media (2012)
11. Zichermann, G.: Fun is the Future: Mastering Gamification (2010)
12. Deterding, S., Sicart, M., Nacke, L., O'Hara, K., Dixon, D.: Gamification. Using Game-design Elements in Non-gaming Contexts. In: CHI 2011 Extended Abstracts on Human Factors in Computing Systems, pp. 2425–2428. ACM, New York (2011)
13. Kapp, K.M.: The Gamification of Learning and Instruction. Pfeiffer (2012)
14. Kumar, J.M., Herger, M.: Gamification at Work: Designing Engaging Business Software. The Interaction Design Foundation (2013) ISBN: 978-87-92964-06-9
15. Peters, D.: Interface Design for Learning: Design Strategies for Learning Experiences. New Riders (2013)
16. Nielsen, J.: 10 Usability Heuristics for User Interface Design, http://www.nngroup.com/articles/ten-usability-heuristics/ (retrieved)
17. Blum, R.: Linux Command Line and Shell Scripting Bible. Wiley Publishing (2008)
18. Norman, D.A.: Emotional Design. Basic Books (2004)
19. Hattie, J.: Visible Learning Meta-Study. Routledge (2009)

# Personalizing E-Learning 2.0 Using Recommendations

Martina Holenko Dlab[1], Natasa Hoic-Bozic[1], and Jasminka Mezak[2]

[1] Department of Informatics, University of Rijeka, Rijeka, Croatia
[2] Faculty of Teacher Education, University of Rijeka, Rijeka, Croatia
{mholenko,natasah}@inf.uniri.hr,
jasminka@ufri.hr

**Abstract.** Recommender systems support users in accessing information available on the Web. This process ensures personalization since recommendations are generated according to user's characteristics. In the educational domain, in the most cases, recommendations refer to learning materials. Besides that, there is a potential for using recommendation techniques in order to personalize other aspects of e-learning context. This paper describes a recommendation model for providing personalization of a collaborative learning process. Well-known recommendation techniques are adapted for online learning environment that consists of an LMS and different Web 2.0 tools. The recommendations are used to support students before and during e-tivities and include four different types of items: optional e-tivities, collaborators, Web 2.0 tools and advice.

**Keywords:** Recommender System, Recommendation Techniques, E-Learning, Collaborative Learning, TEL, Web 2.0.

## 1 Introduction

Recommending items that are potentially useful for the target user or that are within the scope of his/her interests can provide the solution for information overload problem [1]. The usefulness of items (utility) is in recommender systems expressed as a numerical value (rating). This value is determined by the user or it can be predicted. The recommendation problem comes down to the prediction of the unknown utility values in order to recommend item or items with the highest utility to the target user. Recommendation techniques vary depending on the prediction method and can be divided into three main groups [1], [2]: collaborative filtering, content-based and knowledge-based techniques. Hybrid recommenders combine these techniques.

Recommender systems are increasingly used in e-learning [2]. Their advantages also enable personalization within the so-called e-learning 2.0 [3]. E-learning 2.0 emphasizes collaborative learning through a variety of e-learning activities (e-tivities) [4] like online discussions [5], mental mapping, WebQuests. E-learning 2.0 is supported with Web 2.0 tools (e.g., Blogger, Flickr, and YouTube) [4]. Thus with the functionalities of a particular learning management system (LMS), which are the same for all users, students in e-learning 2.0 can use the appropriate third-party

T. Di Mascio et al. (eds.), *Methodologies and Intelligent Systems for Technology Enhanced Learning*, Advances in Intelligent Systems and Computing 292,
DOI: 10.1007/978-3-319-07698-0_4, © Springer International Publishing Switzerland 2014

services available on the Web [6]. Since the choice of recommendation technique depends on the domain, it should be conducted in accordance with the particularities of the context [2].

The paper presents original model of using recommendation techniques as a prerequisite for the development of the system that will personalize the e-learning 2.0. The paper is structured as follows. The second chapter gives overview of items, users and recommendation techniques for the given context. The third chapter describes methods for using selected recommendation techniques for recommendation of four different items: e-tivities, collaborators, tools and advice. The fourth chapter presents conclusions and plans for future work.

## 2    Educational Recommender Systems

Recommendations for education should be distinguished from those for commercial purposes. The aim of educational recommender systems (ERS) is to ensure efficient use of available resources and to support the learning process based on specific learning strategies and pedagogical principles [2]. Domain particularities can be considered in relation to *what* is recommended, *to whom* is recommended and *how* is recommended. Therefore, the identification of potential items, target users and techniques was performed as the initial phase of our recommender system development.

### 2.1    Items

The process of e-learning can be observed as a sequence of actions (activities) performed by students in response to a task. The recommender system endeavors to intelligently recommend a particular action to the student so the variety of items depends on what kind of actions can be recommended [7]. Existing ERS, overviewed in [2], in most cases recommend teaching materials or courses in general [7]. The remaining related work includes recommendations of actions that support the process of learning programming: programming tasks of varying complexity [8], keywords for tagging learning materials [9] and actions with warnings regarding the most common mistakes [10]. TORMES system [7] represents domain independent approach of recommending different actions in dotLRN LMS.

The characteristic of e-learning 2.0 is collaborative learning through e-tivities that are realized with Web 2.0 tools. Thus, we are developing ERS that will enable personalization in environment that includes LMS and a dozen of Web 2.0 tools. Students use the LMS for studying the lessons, solving the (self)assessment tests, and communicating with teachers and colleagues. They use the Web 2.0 tools for the realization of individual or group-based e-tivities (such as writing learning journal with tool Blogger) [11]. In such context, actions that could be supported with recommendations are selection of collaborators for group e-tivities or a specific tool for its realization. Recommended action can also be the participation in an optional e-tivity, for example to collect extra points for the course. In addition, recommendations may be presented in the form of advice to support students (groups) in the e-tivities realization. Accordingly, the selected items that will be tailored to student's characteristics are [11]: optional e-tivities, collaborators (colleague students), Web 2.0 tools, advice.

The prerequisite for mentioned recommendations is that the teacher allows a certain level of flexibility when planning course activities. This involves enabling students to group themselves, planning e-tivities that can be realized with different tools or optional e-tivities between students (groups) will choose one. An example of activity sequence is shown in Fig.1. After introductory *f2f* class, students study lessons, solve online test for self-assessment and participate in WebQuest e-tivity, using one of the offered tools (Blogger, Wikispaces or Google Drive) and divided in groups. They summarize their WebQuest results in a form of presentation published on SlideShare. These activities are followed by one of the optional activities through which students can repeat the main knowledge concepts by making notes, mind mapping or by bookmarking additional web resources before the final online test.

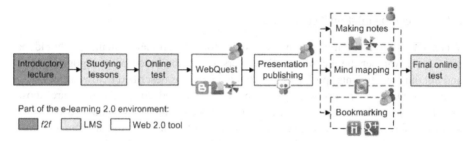

**Fig. 1.** Example of the course workflow that enables different types of recommendations

## 2.2    Target Users

The user of an education recommender system is a student. Recommendation process is based on data about his/her previous actions and achievements, and data about students like him/her. Therefore, specific domain requirements are related to the characteristics which will represent students. Unlike commercial recommender systems where recommendations are based on what users like (interests), in ERS items that students like are not always pedagogically most appropriate for them. Thus, it is often necessary to recommend different items to the students with the same interests [2].

Student's characteristics are represented with student model [12]. Besides interests and preferences, these models can include data regarding knowledge level [9], communication level [7], learning styles [9] and affective states [13]. Needed data can be collected explicitly, using feedback from users, or implicitly (automatically) by collecting data about user interaction with the e-learning environment and recommender system. Priority should be given to implicit collection because it does not increase students' cognitive load [1].

Student model for the ERS that we are developing includes learning styles preferences according to the VARK model [14] and preferences of Web 2.0 tools, both collected via questionnaires at the beginning of the course. The model also contains information about the knowledge level identified on the basis of (self)assessments. An important characteristic is also the activity level which is calculated based on automatically collected data about the students' interaction with Web 2.0 tools. It is calculated periodically during the e-tivities (at intervals specified by the teacher) [11]. The mentioned set of data allows generation of the desired recommendations.

In the context of collaborative learning there is also a need for group recommendations which can be generated based on data from group model or aggregation of data about group members from the student model [12]. Group model for our ERS contains data regarding group activity level. Recommendations based on the other characteristics will be generated using appropriate data from the student model.

## 2.3    Recommendation Techniques

When choosing recommendation techniques for educational domain, it should be considered whether the technique allows personalization based on pedagogical rules, and not only on students' preferences. All techniques, as described below, allow so.

In *collaborative filtering* (CF), items recommended to target user are those preferred by similar students [1]. Similarity between students is calculated based on known preferences. This technique can be used in different learning environments and for recommendations of different items. The filtering can be also done in respect to students' characteristics, which enables the implementation of pedagogical rules (*attribute-based collaborative filtering* method) [2].

*Content-based recommendations* (CB) predict item's usefulness based on the usefulness of the similar items for the target user. Prediction can be based on known preferences (*case-based*) or, more valuable for e-learning, on student's characteristics (*attribute-based*). The later allows the definition of pedagogical rules as part of a recommended strategy but requires (detailed) items representation [2].

Recommendation can be generated based on series of rules as well. Such *knowledge-based systems* (KB) enable recommendations based on expert's (teacher's) knowledge are can be valuable when there is no sufficient amount of data about the student. When it becomes available, collaborative filtering can be used.

*Hybrid recommenders* combine mentioned techniques and, according to [2], often provide the most accurate recommendations because they can overcome problems that occur in a particular technique. Between them cold-start problem should be pointed out. It implies that there is not enough information about the user or the items to provide recommendations [1].

# 3    Recommendations for E-Learning 2.0

This chapter describes our own model of using recommendation techniques in the context of e-learning 2.0, where a set of recommended items includes items insufficiently present in existing educational recommender systems: optional e-tivities, collaborators, Web 2.0 tools, and advice. Table 1 shows the target users and selected recommendation techniques, as well as student's characteristics that will be used in the recommendation process.

**Table 1.** Target users, user's characteristics and selected recommendation techniques

|  |  | *Optional e-tivities* | *Collabora tors* | *Web 2.0 tools* | *Advice* |
|---|---|:---:|:---:|:---:|:---:|
| **Target users** | Student | + | + | + | + |
|  | Group | + | - | + | + |
| **User's characteristics** | Learning style | + | + | + | - |
|  | Tools preferences | + | + | + | - |
|  | Knowledge Level | + | + | - | - |
|  | Activity Level | + | + | - | + |
| **Selected techniques** | Collaborative filtering | - | - | + | - |
|  | Content-based | + | + | + | - |
|  | Knowledge-based | - | - | - | + |

## 3.1    Optional e-tivities Recommendations

This type of recommendation will support students and groups in choosing optional e-tivities. The aim is to rank possible e-tivities for the target student (group) taking into account teacher's criteria.

The chosen technique for this task is *content-based recommendations*, more specifically *attribute-based recommendations*. In general, with this technique, characteristics of items recommended to the target user correspond to his/her needs, which is calculated based on the similarity metrics [2]. Therefore, the similarity of characteristics which represents students (groups) and e-tivities will be calculated and will represent usefulness measure. The teacher, according to pedagogical principles, will define a set of characteristics for calculating the similarity. For example, the teacher may decide that the e-tivities will be recommended depending on the combination of knowledge level of a specific lesson and preferences of learning styles. On the other hand, optional e-tivities can be recommended depending on the activity level of preceding e-tivity or preferences for the tools offered for its realization.

This technique allows assigning weights to characteristics used for similarity calculation that enables the teacher to determine to what extent will each characteristics affect the usefulness of e-tivities. Usefulness calculation based on the characteristics from the student model can be made for the first e-tivity, assuming that students solve VARK questionnaire and specify few Web 2.0 tools preferences at the beginning of the course. In other words, the so-called cold-start problem [1] will not occur.

### 3.2    Collaborators Recommendations

Collaborators recommendations will support students in the selection of collaborators for group-based e-tivities. The aim is to rank the potential collaborators (colleague students) who are, according to the criteria defined by the teacher, the most appropriate for the target student. The same technique as for recommending e-tivities is chosen: *content-based (case-based) recommendations*. Usefulness of potential collaborator will be determined based on the similarity of his/her characteristics with the characteristics of the target student. The teacher will chose the set of characteristics for calculating similarity between the students. The possibility of assigning weights will in this case as well allow him to determine the extent to which each characteristic will affect the final usefulness value. The chosen technique enables the recommendations according to different grouping methods. In the case that students should form homogeneous groups, the colleagues whose characteristics largely coincide with target student's characteristics will be recommended (most useful are the most similar students). On the other hand, a heterogeneous group forming can be encouraged by recommending colleagues with (mutually) different characteristics.

Assuming that students solve VARK questionnaire and specify few tools preferences at the beginning of the course, cold-start problem for new item will not occur. Therefore, usefulness of potential collaborators based on these characteristics can be calculated for the first e-tivity. A possible problem is the lack of diversity [2] in situations when the same collaborators students are recommended for several e-tivities. Since the recommendation criteria is not necessary the same for all e-tivities within the course, this is not considered as major shortcoming. In addition, student's characteristics change over time, which to some extent also affects diversity.

### 3.3    Tools Recommendations

The recommendation techniques will be also used to support selection between the Web 2.0 tools. The aim is to rank the tools offered for an e-tivity in accordance to what student (group) prefer. Therefore, the usefulness of each tool for the target student (group) will be determined based on the Web 2.0 tools preferences.

The *hybrid approach* [2] is chosen, taking into account that the number of items (tools) is relatively small and that the number of students will increase before a number of tools. Therefore, the recommender will switch between collaborative filtering and content-based recommendation based on the number of known student preferences. *Collaborative filtering* technique will be used to solve the cold-start problem for a new student (student whose preferences are not known). Similarity between student will be calculated based on the student characteristics (*attribute-based collaborative filtering* method) [2], namely learning styles preferences. Two students will be considered similar if they have similar learning styles preferences according to the VARK model. To solve cold-start problem for the new tool (tool for which there is no known preferences), the *content-based (case-based) recommendations* will be used. That assumes that target student will like tools that are similar to those he/she prefers (tools similarity will be calculated based on his/her preferences for the other tools) [2].

The list of tools offered for the realization of e-tivity will be presented to students, ranked by usefulness. This does not restrict recommendations to the set of the most popular items, which is a limitation of the collaborative filtering approach. In addition, it allows the student to explore tools that he/she has not used before. The problem of the small number of preferences (*sparse rating problem*) occurs for both selected techniques [2]. In order to overcome this limitation, explicit collection of data regarding tools preferences and collaborative filtering based on the student characteristics is planned.

### 3.4    Providing Advice

Providing advice will be used to motivate students and groups for active participation during the e-tivities (at the end of the intervals defined by the teacher). The aim is to encourage students in reaching higher activity levels which can potentially contribute to greater success in solving the given task. Recommendations will be presented in the form of advice that will relate to different aspects of active participation such as number of different kinds of contributions (e.g. publication of content, commenting, tagging), continuous participation, encouraging collaborators (group members) to participate, etc.

The chosen technique is *knowledge-based recommendations* [1]. Using selected method, recommended items are associated with the student's (group's) needs based on explicitly stated "if...then..." expert rules. Target student's characteristics will be compared the to the teacher expectations, so the rules will contain a number of control parameters. According to that, this kind of recommendations will greatly depend on the pedagogical principles derived from expert's (teacher's) knowledge. Example of advice might be: "*Your activity level for [e-tivity_title] is not satisfying. E-activity lasts till [end_date] so it is highly recommended that you participate to a greater extent.*". It should be noted that the lack of this approach is the complexity of formal representation of the expert knowledge.

## 4    Preliminary Results, Conclusions and Future Work

Besides learning materials, there are others items in the context of e-learning 2.0 that can be adapted to students' characteristics. For the presented recommendation model optional e-tivities, collaborators, Web 2.0 tools and advice were pointed out. Student's characteristics that have potential to ensure personalization and overcome possible problems (i.e. cold-start) were identified. Recommendation techniques and methods were selected in accordance with the structure of the e-learning environment and available students' data. The recommendation model described in this paper was implemented in the prototype of E-Learning Activities Recommender System - ELARS [15]. The system was used for two e-courses at the Department of Informatics, University of Rijeka, Croatia. With the help of the e-learning designer familiar with the model and the system's authoring component, teachers planned the e-tivities and adjusted the recommendation criteria according to the desired pedagogic strategies.

The most interesting findings of the survey performed with students (N=42) are that the system positively influenced on their level of engagement in e-tivities (74%, while 17% was neutral) and their motivation for learning (52%, while 36% was neutral). They were satisfied with received recommendations (50%, while 32% was neutral), find ELARS useful for the context of e-tivities (57,4%, while 30,1% was neutral) and easy to use (87%, while 13% was neutral).

These preliminary results are encouraging for further work on the system's e-tivities authoring component. Since three out of four types of recommendations depend on teacher's knowledge, we aim to develop an user friendly interface which will enable teachers to independently plan e-tivities workflows and adjust the personalization methods. This will be followed by evaluation of system's usefulness and usability from teacher's perspective in order to get insights to possible improvements of the recommendation model and e-tivities authoring component.

**Acknowledgments.** The research has been conducted under the project "E-learning Recommender System" (reference number 13.13.1.3.05) supported by University of Rijeka (Croatia).

# References

1. Adomavicius, G., Tuzhilin, A.: Toward the next generation of recommender systems: A survey of the state-of-the-art and possible extensions. IEEE Educational Activities Department (2005)
2. Manouselis, N., Drachsler, H., Vuorikari, R., Hummel, H., Koper, R.: Recommender Systems in Technology Enhanced Learning. In: Ricci, F., Rokach, L., Shapira, B., Kantor, P.B. (eds.) Recommender Systems Handbook, pp. 387–415. Springer, Boston (2011)
3. Downes, S.: E-learning 2.0 (2005), http://elearnmag.acm.org/featured.cfm?aid=1104968 (accessed January 9, 2014)
4. Salmon, G.: E-tivities: the key to active online learning. Comput. Educ. 45, 223 (2002)
5. Holenko Dlab, M., Hoić-Božić, N.: Using Online Discussions in a Blended Learning Course (2008), http://online-journals.org/i-jet/article/view/630/594
6. Holenko Dlab, M., Hoić-Božić, N.: An Approach to Adaptivity and Collaboration Support in a Web-Based Learning Environment (2009), http://online-journals.org/i-jet/article/view/1071
7. Santos, O.C., Boticario, J.G.: Modeling recommendations for the educational domain. Procedia Comput. Sci. 1, 2793–2800 (2010)
8. Michlík, P., Bieliková, M.: Exercises recommending for limited time learning. Procedia Comput. Sci. 1, 2821–2828 (2010)
9. Klasnja-Milicevic, A., Vesin, B., Ivanovic, M., Budimac, Z.: E-Learning personalization based on hybrid recommendation strategy and learning style identification. Comput. Educ. 56, 885–899 (2010)

10. Sheth, S., Arora, N., Murphy, C., Kaiser, G.: Wehelp: A Reference Architecture for Social Recommender Systems. Architecture, 46–47 (2010)
11. Holenko Dlab, M., Hoić-Božić, N.: Recommender System for Web 2.0 Supported eLearning. In: IEEE Global Engineering Education Conference Proceedings (2014)
12. Brusilovsky, P., Millán, E.: User Models for Adaptive Hypermedia and Adaptive Educational Systems. Adapt. Web Methods Strateg. Web Pers. 4321, 3–53 (2007)
13. Leony, D., Pardo, A., Parada, G.: H.A., Delgado Kloos, C.: A cloud-based architecture for an affective recommender system of learning resources. In: International Workshop on Cloud Educational Environments, pp. 1–2 (2012)
14. Fleming, N.D.: I'm different; not dumb. Modes of presentation (VARK) in the tertiary classroom. In: Zelmer, A. (ed.) Research and Development in Higher Education, Proceedings of the Annual Conference of the Higher Education and Research Development Society of Australasi (HERDSA), pp. 308–313 (1995)
15. ELARS - E-Learning Activities Recommender System (demo home page), http://161.53.18.114/elarsdemo (accessed January 9, 2014)

# The Assessment of Motivation in Technology Based Learning Environments: The Italian Version of the Achievement Goal Questionnaire-Revised

Daniela Raccanello, Margherita Brondino, Margherita Pasini, and Bianca De Bernardi

Department of Philosophy, Education and Psychology, University of Verona (Italy),
Lungadige Porta Vittoria 17, 37129 Verona, Italy
{daniela.raccanello,margherita.brondino,margherita.pasini,
bianca.debernardi}@univr.it

**Abstract.** The increased use of technology within the educational field gives rise to the need of developing valid instruments enabling to measure key constructs in fast ways, often involving many people. Therefore, this work explored some psychometric properties–in terms of factor structure and alpha coefficients–of a first Italian version of the Achievement Goal Questionnaire-Revised [8], as a first step preceding a future computerized implementation of the instrument. Seventy-seven university students completed the questionnaire, referring to a specific course. Each questionnaire included 12 items, three for each goal: Mastery-approach, mastery-avoidance, performance-approach, and performance-avoidance goals. Confirmatory factor analyses indicated the goodness of the hypothesized model, and its superiority compared to several alternative models; also alpha coefficients resulted adequate. These findings supported the validity of the adapted instrument, and were discussed considering possible future uses to assess motivational outcomes related to learning tasks within computer-based environments.

**Keywords:** Achievement Goals, Scale Validation, motivation for technology-based learning environments.

## 1    Introduction

Within educational psychology, the relevance of motivation in learning contexts has been largely addressed, with an increasing body of knowledge revealing its complex links with cognition, affect, and behaviour [13, 19]. Among different motivational constructs, a key role is played by achievement goals [10], as "cognitive–dynamic aims that focus on competence" [8]. Achievement goals' assessment can be considered a way to evaluate learning motivation. At the same time, different achievement goals seem to affect students' preference for the learning environments. This suggests the importance of achievement goals' assessment also for the definition of technology based learning environments, adaptive to the students' motivational profiles. Achievement goals comprise two dimensions: Definition–in terms of mastery and performance constructs–and valence–in terms of approach or avoidance

T. Di Mascio et al. (eds.), *Methodologies and Intelligent Systems for Technology Enhanced Learning*, Advances in Intelligent Systems and Computing 292,
DOI: 10.1007/978-3-319-07698-0_5, © Springer International Publishing Switzerland 2014

strivings. The first dimension, identified back in the eighties [5], refers to criteria for judging competence: for mastery goals individuals strive to reach competence, while for performance goals individuals focus on comparisons with others [2, 5, 16]. According to the first empirical results, mastery goals had adaptive consequences for learning, while performance goals were associated to maladaptive behaviours [5]. However, these negative consequences (not so univocally found compared to positive consequences of mastery goals) have been recently revised, after the identification of the approach-avoidance dimension that can characterize both performance goals and mastery goals [6, 7]. Specifically, individuals can compare themselves to others in two ways: Striving either to demonstrate competence (approach goals) or not to demonstrate incompetence (avoidance goals). As a result, a 2 X 2 model comprising four kinds of goals has been proposed: Mastery-approach goals, mastery-avoidance goals, performance-approach goals, and performance-avoidance goals. The introduction of this distinction has cleared up some previous ambiguous results, according to which, for example, performance goals were linked both to positive aspects, such as high self-efficacy and effort, and negative effects, such as low task persistence [11, 18, 20]. Negative consequences were probably linked to performance-avoidance goals, and positive consequences to performance-approach goals [20], which frequently exhibit, however, high positive correlations [13]. Concerning the most recent developments of achievement goals models, it is worth noting that until now scarce attention has been paid to mastery-avoidance goals compared to the other three goals, and that a further distinction about mastery goals has also been proposed, distinguishing a task- and a self-component [9].

One of the instruments currently most used to evaluate achievement goals [15] is the Achievement Goal Questionnaire, AGQ [7], which operationalizes the above mentioned definition allowing to asses the four goals described in the 2 X 2 model. However, the authors of the AGQ have recently identified some conceptual and methodological problems in their instrument, such as references to values, concerns, or affect rather than goals; lack of separation between aims and underlying reasons; absence of consistency in item content. To solve these problems, they proposed the Achievement Goal Questionnaire-Revised, AGQ-R [8], first tested with American college undergraduates about their first psychology exam [8]. The analyses indicated the structural validity of the AGQ-R: The model distinguishing the four factors was confirmed, and it was characterized by better fit indexes compared to alternative models. Also its predictive validity was supported, considering both antecedents such as need for achievement and fear of failure, and consequences such as intrinsic motivation and achievement.

The increased use of technology within the educational field gives rise to the need of developing valid instruments enabling to measure key constructs in fast ways, often involving many people. On the one hand, the key characteristics of the AGQ-R make it a good product also for technology based environment, in which features such as accuracy in the correspondence between the measured concepts (i.e., achievement goals) and operationalized items is essential. Also other features of the AGQ-R guarantee the goodness of its generalizability to computer-based contexts, such as brevity (it is formed by only 12 items), word simplicity, and answer modality (Likert-type scale). On the other hand, proposing this instrument through computerized means, for example e-mailing an URL to the participants and enabling to respond to the questionnaire with an online methodology (similarly for example to Meier and colleagues' procedure [14]), has the multiple advantages to involve a higher number of persons simultaneously, without the constraints of requesting their physical presence in the same place, reducing at the same time the costs of the administration.

Therefore, this work explored some psychometric properties–in terms of factor structure and alpha coefficients–of a first Italian version of the AGQ-R [8], to our knowledge not available yet in the Italian context. This contribution represents a first step preceding a future computerized implementation of the instrument. Following the authors of the original instrument, we hypothesized that the model in which the four achievement goals load on four separate factors had good fit indexes, and we expected that it revealed better compared to alternative three- and two-factors models. As a secondary aim, we expected positive correlations between goals sharing a dimension, and explored mean differences between the four goals.

## 2    Method

### 2.1    Participants

A total of 77 university students (70 females), ranging in age from 19 years and 5 months to 43 years and 11 months, with a mean age of 22 years and 10 months, participated. They came from a variety of socio-economic backgrounds. They were enrolled in a Faculty of Science of Education in Northern Italy, and were attending the first year (n = 2; 3%), the second year (n = 40; 52%), the third year (n = 32; 41%), or subsequent years (n = 3; 4%). They were tested during a psychology (n = 46; 60%) or a sociology class (n = 31; 40%). Preliminary analyses revealed no differences according to year or class, not subsequently considered.

### 2.2    Material and Procedure

Participants were administered a written questionnaire on a voluntary basis. They were assured about anonymity and that their answers would not influence course evaluations. Each session lasted about ten minutes.

#### 2.2.1    Achievement Goal Questionnaire-Revised (AGQ-R)

The questionnaire included an Italian adaptation of the AGQ-R [8], translated from English to Italian and then from Italian to English according to the back-translation procedure, with the permission of A. J. Elliot. The Italian version of the AGQ-R included 12 items referred to on of the courses that students were attending during that class, specifically psychology or sociology courses. Three items regarded mastery-approach goals (e.g., "My aim is to completely master the material presented in this class"), three items mastery-avoidance goals (e.g., "My goal is to avoid learning less than it is possible to learn"), three items performance-approach goals (e.g., "I am striving to do well compared to other students"), and three items performance-avoidance goals (e.g., "My aim is to avoid doing worse than other students"). See Results for alpha-values. The item order was the same proposed by the authors of the instrument [8]. Compared to the original questionnaire, some changes were made in the linguistic formulation as regards the reference to a specific course (e.g., "in this course"): It was proposed both where included by the authors of the instrument, and in some other items (e.g., items 2 and 6) in order to differentiate items regarding the same goal and to make clear the comparison group. It was not included in all the items to avoid repetitiveness in the task. Another change regarded the evaluation scale, which was a 5-point Likert-type scale (1 = *not at all true of me* and 5 = *very true of me*), and not a 7-point Likert-type scale. This choice was based on possible future uses with younger students, for whom it is difficult using response scales with a high number of levels.

### 2.3    Data Analyses

Confirmatory factor analyses (CFA) were performed to evaluate the psychometric properties of the Italian version of the AGQ-R using AMOS 7.0 [3]. A first CFA enabled to test the goodness of the hypothesized model, in which mastery-approach items, mastery-avoidance items, performance-approach items, and performance-avoidance items load, respectively, on four distinct latent factors. The following indexes were considered to assess the goodness of fit of the model [8]: Chi-square degree of freedom ratio ($\chi^2/df$), comparative fit index (CFI), incremental fit index (IFI), root-mean-square error of approximation (RMSEA), Akaike information criterion (AIC), and Bayesian information criterion (BIC). The next threshold values were taken into account: $\chi^2/df \leq 2$, CFI and IFI $\geq .90$, and RMSEA $\leq .08$ [4].

The same indexes and threshold values were used to compare the hypothesized model to other six models, specifically: (a) a trichotomous model A, corresponding to the trichotomous achievement goal model [7], in which the mastery-approach items and the mastery-avoidance items load on the same latent factor, while the performance-approach items and the performance-avoidance items load on other two separate latent factors; (b) a trichotomous model B, with performance items loading together, and mastery-approach and mastery-avoidance items loading on separate factors; (c) a trichotomous model C, with avoidance items loading together, and mastery-approach and performance-approach items loading on separate factors; (d) a trichotomous model D, with approach items loading together, and mastery-avoidance and performance-avoidance items loading on separate factors; (e) a mastery-performance model, with mastery items loading together, and performance items loading together; and (f) a approach-avoidance model, with approach items loading together, and avoidance items loading together.

Finally, $\alpha$-coefficients for items about the same goals were calculated to evaluate internal consistency; descriptive statistics and intercorrelations were presented; $t$-tests were run to compare the goals. The level of significance was $p < .05$.

# 3    Results

## 3.1    Confirmatory Factor Analysis and Internal Consistency

The results from the CFA run on the hypothesized model were good enough to support the model. Factor loadings were all statistically significant and fit indexes were quite high (see Figure 1 for the model; see Table 1 for overall fit indexes).

Concerning achievement goals, all α-values were higher than .70 (mastery-approach goals: α = .77; mastery-avoidance goals: α = .71; performance-approach goals: α = .88; performance-avoidance goals: α = .80), minimum value recommended in the literature [17] to consider internal consistency as adequate.

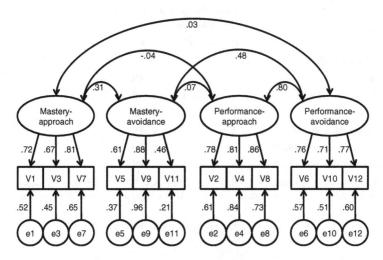

**Fig. 1.** Factorial structure of the Italian version of the AGQ-R

## 3.2    Comparisons with Alternative Models

None of the alternative models showed acceptable fit to the data (see Table 1 for fit indexes). All the results of the alternative models were worst as confirmed by the Log-likelihood ratio tests that pointed out how the hypothesized model provided better indexes than any alternative models.

## 3.3    Descriptive Statistics, t-tests, and Intercorrelations

Students' responses on the items about the four factors were averaged together. Descriptive statistics and intercorrelations are presented in Table 2.

**Table 1.** Fit indexes and comparisons between the 2 X 2 model and alternative models

| Variable | $\chi^2/df$ | CFI | IFI | RMSEA | AIC | BIC |
|---|---|---|---|---|---|---|
| | | | Fit indexes | | | |
| Hypothesized model | 1.76 | .91 | .91 | .10 | 144.63 | 214.95 |
| Trichotomous model A | 2.72 | .79 | .79 | .15 | 192.59 | 255.87 |
| Trichotomous model B | 2.34 | .83 | .84 | .13 | 173.07 | 236.35 |
| Trichotomous model C | 2.83 | .77 | .78 | .16 | 198.10 | 261.38 |
| Trichotomous model D | 5.17 | .48 | .50 | .23 | 317.58 | 380.86 |
| Mastery-performance model | 2.91 | .75 | .76 | .16 | 203.98 | 262.58 |
| Approach-avoidance model | 5.79 | .38 | .40 | .25 | 357.10 | 415.69 |

| | Log-likelihood ratio test (model comparison) | | |
|---|---|---|---|
| | $df$ | $\chi^2$ | $p$ |
| Hypothesized model versus | | | |
| Trichotomous model A | 3 | 53.96 | < .001 |
| Trichotomous model B | 3 | 34.44 | < .001 |
| Trichotomous model C | 3 | 59.47 | < .001 |
| Trichotomous model D | 3 | 178.95 | < .001 |
| Mastery-performance model | 5 | 69.35 | < .001 |
| Approach-avoidance model | 5 | 222.47 | < .001 |

*Note.* CFI = comparative fit index; IFI = incremental fit index; RMSEA = root-mean-square error of approximation; AIC = Akaike information criterion; BIC = Bayesian information criterion.

**Table 2.** Descriptive statistics and intercorrelations

| Variable | $M$ (SD) | 1 | 2 | 3 | 4 |
|---|---|---|---|---|---|
| 1. Mastery-approach goals | 4.12 (.68) | - | | | |
| 2. Mastery-avoidance goals | 3.24 (.98) | .26* | - | | |
| 3. Performance-approach goals | 2.17 (.94) | -.03 | .05 | - | |
| 4. Performance- avoidance goals | 2.26 (.95) | .01 | .37** | .67*** | - |

*$p < .05$, **$p < .01$, ***$p < .001$.

Regarding achievement goals, paired-sample $t$-tests revealed higher scores for mastery-approach goals than mastery-avoidance goals, $t(76) = 7.49$, $p < .001$, which in turn had higher scores than both performance-approach goals, $t(76) = 7.03$, $p < .001$, and performance-avoidance goals, $t(76) = 7.90$, $p < .001$. The two performance goals did not differ one another.

Concerning intercorrelations, significant positive associations emerged between mastery-approach and mastery-avoidance goals; performance-approach and performance-avoidance goals; and mastery-avoidance and performance-avoidance goals. Differently from Elliot and Murayama [8], neither mastery-approach and performance-approach goals, nor mastery-avoidance and performance-approach goals were positively correlated; however, in their work the level of significance associated with these correlations was higher compared to that associated to the previous correlations.

## 4    Discussion and Conclusions

Nowadays, the role of Technology Enhanced Learning (TEL) is assuming an increasing role within the educational field, where the mutual links between cognitive, motivational, and affective processes have been clearly documented [12, 19]. The very frequent use of technology in learning contexts gives rise to the need of developing methodologies enabling to measure key constructs in fast ways, often involving many people. This possibility is guaranteed by the availability of valid instruments adapted to different contexts.

In this line, the main aim of this study was to test a first Italian version of the AGQ-R [8] with university students, as a first step for its use to assess motivational constructs in technology based environments. Confirmative factor analyses indicated the goodness of the hypothesized model, and its superiority compared to several alternative models; also alpha coefficients resulted adequate. A second step of this research study is to create an on-line version of the instrument, using the software ApsymSurvey, a modified version of the core of LimeSurvey (http://www. limesurvey.org) [1]. This could be used in future studies to monitor students' achievement goals during the whole university period, with all the advantages of an online methodology [e.g., 14].

Notwithstanding possible limitations related to the use of self-report methods, the AGQ-R could be utilized in a great variety of ways to evaluate key constructs related to learning. Indeed, this questionnaire has the potential to be used worthwhile in TEL contexts, to evaluate individuals' motivation towards school or university learning but also towards a wider range of different learning programs within computer-based environments.

## References

1. Apsym Lab, ApsymSurvey online survey software. Verona University, Italy (2013)
2. Ames, C.: Classrooms: Goals, structures, and student motivation. Journal of Educational Psychology 84, 261–271 (1992)
3. Arbuckle, J.L.: Amos (Version 7.0) (computer program). SPSS, Chicago (2006)
4. Browne, M.W., Cudeck, R.: Alternative ways of assessing model fit. In: Bollen, K.A., Long, J.S. (eds.) Testing Structural Equation Models, pp. 136–162. Sage, Beverly (1993)
5. Dweck, C.S., Leggett, E.L.: A social-cognitive approach to motivation and personality. Psychological Review 95, 256–273 (1988)
6. Elliot, A.J.: Approach and avoidance motivation and achievement goals. Educational Psychologist 34, 169–189 (1999)
7. Elliot, A.J., McGregor, H.: A 2 X 2 achievement goal framework. Journal of Personality and Social Psychology 80, 501–519 (2001)
8. Elliot, A.J., Murayama, K.: On the measurement of achievement goals: Critique, illustration, and application. Journal of Educational Psychology 100, 613–628 (2008)

9. Elliot, A.J., Murayama, K., Pekrun, R.: A 3 X 2 achievement goal model. Journal of Educational Psychology 103, 632–648 (2011)
10. Gegenfurtner, A., Hagenauer, G.: Achievement goals and achievement goal orientations in education. International Journal of Educational Research 61, 1–4 (2013)
11. Harackiewicz, J.M., Durik, A.M., Barron, K.E.: Multiple goals, optimal motivation, and the development of interest. In: Forgas, J.P., Williams, K.D., Laham, S.M. (eds.) Social Motivation: Conscious and Unconscious Processes, pp. 21–39. Cambridge University Press, New York (2005)
12. Hulleman, C.S., Schrager, S.M., Bodmann, S.M., Harackiewicz, J.M.: A meta-analytic review of achievement goal measures: Different labels for the same constructs or different constructs with similar labels? Psychological Bulletin 136, 422–449 (2010)
13. Law, W., Elliot, A.J., Murayama, K.: Perceived competence moderated the relation between performance-approach and performance-avoidance goals. Journal of Educational Psychology 104, 806–819 (2012)
14. Meier, A.M., Reindl, M., Grassinger, R., Berner, V.D., Dresel, M.: Development of achievement goals across the transition out of secondary school. International Journal of Educational Research 61, 15–25 (2013)
15. Muis, K.R., Winne, P.H., Edwards, O.V.: Modern psychometrics for assessing achievement goal orientation: A Rasch analysis. British Journal of Educational Psychology 79, 547–576 (2009)
16. Nicholls, J.G.: Achievement motivation: Conceptions of ability, subjective experience, task choice, and performance. Psychological Review 91, 328–346 (1984)
17. Nunnally, J.C., Bernstein, I.H.: Psychometric Theory, 3rd edn. McGraw-Hill, New York (1994)
18. Pajares, F., Britner, S.L., Valiante, G.: Relation between achievement goals and self-beliefs of middle school students in writing and sciences. Contemporary Educational Psychology 25, 406–422 (2000)
19. Pintrich, P.R.: A motivational science perspective on the role of student motivation in learning and teaching contexts. Journal of Educational Psychology 95, 667–686 (2003)
20. Urdan, T.: Predictors of academic self-handicapping and achievement: Examining achievement goals, classroom goal structures, and culture. Journal of Educational Psychology 96, 251–264 (2004)

# Development of Sign Language Communication Skill on Children through Augmented Reality and the MuCy Model

Jonathan Cadeñanes and Angélica González Arrieta

Department of Computer Science and Automation, Faculty of Science,
University of Salamanca, Spain
{cadenanes,angelica}@usal.es

**Abstract.** This paper shows a Sign Language Teaching Model (SLTM) called: Multi-language Cycle for Sign Language Understanding (MuCy). It serves as complementary pedagogical resource for Sign Language (SL) teaching. A pilot lesson with the Rainbow Colors was conducted at the Association of Parents of Deaf Children of Salamanca in order to determine the Percentage of Development of the Sign Language Communication Skill (SLCS) and others within a Collaborative Learning Environment with Mixed-Reality (CLEMR).

**Keywords:** Augmented Reality, Sign Language Communication Skill, Unity3D, Collaborative Learning Environment, pedagogical tool.

## 1  Introduction

To develop different Communication Skills (CS) on deaf children their individual learning readiness and intellectual capacities have to be taken into account. Teachers from preschool to secondary education sometimes need more creative teaching methods to develop the SLCS on deaf students [18]. And one suitable option for doing this (considering the educational curriculum) is the Augmented Reality (AR) technology, as it includes visual and interactive digital contents which are viewed in the real world.

Since deaf students are visual learners, they need to use images in order to understand the ideas and concepts from their surrounding environment. Teachers turn to consider the use of non-immersive and immersive Interactive Digital Learning Environments (IDLE) since they have shown to be effective methods for teaching Sign Language (SL) [1].

On IDLE two considerations are essential: The first, is a taxonomical framework that classifies the Mixed-Reality (MR). A Reality-Virtuality Continuum establishes the process to change between realities by using interfaces. The second (in accordance with the AR displays) [17] to move between those realities into a smooth transition, users have to employ a monitor-based (non-immersive) or Window on the World (WoW) device adapted for their learning needs.

This paper is organized as follows: In Section 2, we mention the two projects more similar to the one that we are proposing. In Section 3, the SLTM MuCy

T. Di Mascio et al. (eds.), *Methodologies and Intelligent Systems for Technology Enhanced Learning*, Advances in Intelligent Systems and Computing 292,
DOI: 10.1007/978-3-319-07698-0_6, © Springer International Publishing Switzerland 2014

and the Pedagogical Materials (PM) are explained in more detail as well as the criteria used to validate the model. Also we present a SL pilot lesson conducted at the Association of Parents of Deaf Children of Salamanca (ASPAS) [2]. In Section 4, we draw the conclusions. And finally, in section 6. The future research is mentioned.

## 2   Related Works

The MagicBook proposed by Billinghurst et al. [4,5,6,7,8]; is considered the best tool that allows users to move between Reality and Virtuality. This project covers three levels of MR within a Multi-level Collaborative Learning Environment.

The first level is the Reality. The use of an AR book as a tangible interface allows readers to interchange opinions while they are are sharing their learning experience. But if they want to learn on the second level (with the AR system), they have to use another interface (AR displays or PC screens). On this level, users can see digital worlds and avatars projected onto real objects (the physical books). Multiple learners can gather around the screen and still share information in both real and digital worlds.

By pressing a switch button on their AR displays, the users will no longer be in the real world because their view is totally immersive. This is the last level: Virtuality, on which multiple users can experience the AR learning scenes by being connected to workstations. The MagicBook supports collaboration on three levels: As a physical object (the book), as an AR object (avatars) and as an immersive Virtual Space (digital scenes).

The use of Immersive Learning Environments such as Mathsigner™ allows deaf students to interact in real time with digital avatars by making signs through a Glove-based SL input recognition system [1]. As a result students can learn mathematical concepts and American Sign Language (ASL) terminology.

Mathsigner™ can be displayed on immersive systems such as Flex™ and reFex™. For these versions it is necessary to use glove and eye-wear displays in order to enter the system and interact with it. For the non-immersive versions a computer desktop program was created to be used at schools or at homes.

## 3   Multi-language Cycle for Sign Language Understanding, Pedagogical Materials and Pilot Lesson

We propose a Sign Language Teaching Model (SLTM) called: Multi-language Cycle for Sign Language Understanding (MuCy). It considers the diverse educational needs and individual communication development of deaf people from different ages. It also takes into account the fact that Education is considered as a part of social life [11] bearing in mind that deaf people lead social lives as well. It is important to offer them creative solutions for their social integration into a society based on communication. To the extent they learn to use different CS they will be more confident to express their ideas and feelings within diverse socio-cultural groups. Therefore, the model helps to establish (at SL schools) an outline to create a Collaborative Learning Environment with Mixed-Reality (CLEMR).

The theoretical background of the SLTM we are proposing is based on Lev Semionovich Vigotsky's Principles of Social Education for deaf and dumb children [21], as well as the Zone of Proximal Development (ZPD) [11,14] and the Milgram's Reality-Virtuality Continuum or Mixed Reality (MR) [17].

The model's design is supported by the neuropsychological findings that have shown that deaf children can develop good reading and speaking skills by learning these concepts at an early-age [15], as well as the fact that their spoken language development (as the result of reading processes) serves to increase other CS mentioned before. Teachers as the mediators between learning interfaces and students regulate the learning experience through the ZPD and by promoting information exchanges between realities and users [16].

Deaf children can learn by using interfaces and digital content according to their intellectual capacity. Then with the help of others they imitate signs, interact with information and share knowledge within technological and Socio-cultural influences. In brief, the SLCS is enhanced by technology, people, and information (knowledge).

The MuCy model establishes two psycho-motor teaching levels of Education for SL Communication (Fig.1). On the first level of education, the objective is to teach the proper use of signs in relationship with their visual references and their written words to establish a logical connection of meanings between them.

By *signs*, we refer to the standardized group of body movements in a logical sequence that has been established by the educational authorities or professionals in the field. The *visual references* are the words or written sentences that correspond in meaning to the specific signs performed. The *written words* refers to the action of writing down on paper the meaning of that word.

On the second level of education, we present the *verbalization of the written word*. It is believed that deaf people can not speak, but that is not true. The process of reading books at home at an early age as a primary language skill has enabled the development of speech [15]. Furthermore, some deaf children have successfully acquired the ability to read from the face, lip and tongue movements, so there is no need to use SL to communicate [21].

To create the Unity3D AR desktop application for SL, first we used Blender 2.69 for the modeling and animation process [3]. Next, to build the AR scenes which can be displayed on PCs screens, tablets or AR displays we used the Cross-platform Game Engine Unity3D V4.3 [19] and the Vuforia-AR Extension [20].

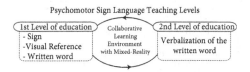

**Fig. 1.** Multi-language Cycle for Sign Language Understanding (MuCy model)

**Fig. 2.** Sign Language Pedagogical Materials (SLPMs) as learning interfaces based on the VR-Continuum. A) Some signs for colors, B) SL Book with sections for reading and writing exercises, C) Avatar making signs on WoW-Videos, D) Vuzix AR display, E) A Unity3D AR avatar making signs.

The use of an FBX format allowed us to view on the screens (in high-definition) the avatar's face and body animations.

The Sign Language Pedagogical Materials (SLPMs) (Fig.2) as learning interfaces presented in tangible and digital formats were designed to be used with the MuCy model in accordance with the RV-Continuum. The aim is to teach concepts to children from a basic to complex levels of understanding (knowledge). They also can choose the interface that better suits their learning needs. The SLPMs developed to be used together with the MuCy model are:

1) *The SL book* as a tangible interface for reading and writing exercises facilitates the adequate understanding of text and images (visual references) that correspond to specific SL positions. With this material, students can train their minds by making associations between signs and their correspondent words or phrases.

2) *The animations* as intermediate interfaces between Reality and digital contents (WoW) allow children learn by imitation while animated avatars perform the appropriate SL positions.

**Table 1.** Likert's scale survey to validate the MuCy model and the SLPMs

| i | Question | Mean | Std.Dev. | % |
|---|----------|------|----------|---|
| Q1 | The SLPMs help deaf children to remember information through memorization. | 4.00 | 1.414 | 80% |
| Q2 | The two educational levels of the MuCy model help deaf students to cognitively understand relevant information from the SL. | 5.00 | .000 | 100% |
| Q3 | Teaching Communication Skills such as reading, writing and speaking help deaf students to create solutions to the socio-cultural problems they face. | 5.00 | .000 | 100% |
| Q4 | Learning with a CLEMR helps deaf students to understand a complex situation in parts in order to create diverse learning solutions. | 5.00 | .000 | 100% |
| Q5 | Learning with interactive technology helps children increase their learning achievement. | 4.50 | .707 | 90% |
| Q6 | I would like to use these pedagogical materials as complementary teaching resources either at home or at school. | 5.00 | .000 | 100% |
| Q7 | The MuCy model helps deaf children to organize their learning process according to their educational needs. | 5.00 | .000 | 100% |
| Q8 | With these pedagogical materials it is easier to explain the SL positions to the children. | 4.00 | 1.414 | 80% |
| Q9 | Learning with AR avatars increases the interest in speech and makes the children feel more confident that they will learn to speak. | 5.00 | .000 | 100% |
| Q10 | The SL book is an adequate tool for teaching the reading and writing for an specific topic. | 4.50 | .707 | 90% |

3) *The Unity3D AR desktop application for SL* as intermediate interface uses a marker-based tracking system which can be adapted with AR display devices such as Vuzix for immersive learning experiences [22].

In order to validate the MuCy model and the SLPMs we are founded firstly, on the Principles of Learning and Teaching P-12 which are established by the Department of Education and Early Childhood Development [13]. Then, on the Danielson's Group Framework for Teaching [12], and finally on the Bloom's Taxonomy of Educational Objectives [9]. With these references we considered the most relevant aspects of each of them to design a likert's scale survey of five points [Table 1]. The respondents were two teachers from ASPAS.

The topic chosen for the SL pilot lessons is the Rainbow Colors. The SL lesson with the Colors was conducted in order to measure the Percentage of Development of SLCS and other CS reached by three deaf children (Fig.3A). This lesson had a duration of one hour with students located in different classrooms. A six-year-old student (Group A) learnt SL without using the MuCy Model and the SLPMs. The other two students aged six and seven (Group B) attended the lesson together with the materials within a CLEMR (Table 2).

For the Colors lesson we made 16 videos (including the signs for the concepts of light, dark and color). The duration of each video was approximately 6 seconds. For every minute each student watched and imitated an average of 8 to 10 SL positions. The lesson was divided into four activities. Each of them corresponding to a specific SLPMs.

Activity one corresponded to the animated videos. The children watched the avatars performing signs on the Tablet (Case A) and on the PC screen (Case B). All the students had to imitate the SL positions right after the avatars. For Activities two and three, the children had to use the SL book to practice the reading of the words for each color or concept. Next, they had to write those words down on the book. Finally (immediately after the writing exercises) they had to perform the SL positions corresponding to those words.

At the last Activity, the children first had to use the markers printed on the pages to display the animated avatars on the PC screen. Then, teachers taught

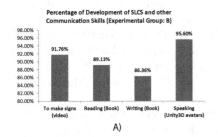
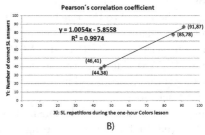

A)                              B)

**Fig. 3.** A) Percentage of Development of SLCS and other Communication Skills of the Experimental Group. B) Correlation analysis between the SL repetitions during the one-hour Colors lesson and the students' correct SL answers.

**Table 2.** SL Repetitions, correct answers and total scores in One-hour lesson. A) Control Group: 1 student, B) Experimental Group:2 students.

**A) Control Group. The Rainbow Colors lesson.**

| Activity | T (mins) | SL Reps. Goal | Session SL Reps. | | | | Xi | Yi | Percent | Score |
|---|---|---|---|---|---|---|---|---|---|---|
| | | | MS | RD | WR | SP | | | | |
| 1 | 20 | 100 | 80 | 0 | 0 | 0 | 80 | 72 | 90.00% | 9.0 |
| 2 | 10 | 50 | 0 | 45 | 0 | 0 | 45 | 35 | 77.78% | 7.8 |
| 3 | 10 | 50 | 0 | 0 | 40 | 0 | 40 | 28 | 70.00% | 7.0 |
| 4 | 20 | 100 | 0 | 0 | 0 | 75 | 75 | 70 | 93.33% | 9.3 |
| **Total** | **60** | **300** | **80** | **45** | **40** | **75** | **240** | **205** | **85.42%** | **8.5** |
| **Mean Value** | 15 | 75 | 20 | 11.25 | 10 | 18.75 | 60 | 51.25 | **82.78%** | **8.28** |
| **Std. Dev** | | | | | | | | | | **1.08** |

**B) Experimental Group. The Rainbow Colors lesson.**

| Activity | T (mins) | SL Reps. Goal | Session SL Reps. | | | | Xi | Yi | Percent | Score |
|---|---|---|---|---|---|---|---|---|---|---|
| | | | MS | RD | WR | SP | | | | |
| 1 | 20 | 100 | 85 | 0 | 0 | 0 | 85 | 78 | 91.76% | 9.2 |
| 2 | 10 | 50 | 0 | 46 | 0 | 0 | 46 | 41 | 89.13% | 8.9 |
| 3 | 10 | 50 | 0 | 0 | 44 | 0 | 44 | 38 | 86.36% | 8.6 |
| 4 | 20 | 100 | 0 | 0 | 0 | 91 | 91 | 87 | 95.60% | 9.6 |
| **Total** | **60** | **300** | **85** | **46** | **44** | **91** | **266** | **244** | **91.73%** | **9.2** |
| **Mean Value** | 15 | 75 | 21 | 12 | 11 | 23 | 66.5 | 61 | **90.72%** | **9.07** |
| **Std. Dev** | | | | | | | | | | **0.39** |

the students to move their lips and tongues to reproduce sounds and to practice speech.

## 4   Conclusions

We have presented in this article a SLTM called MuCy. It established two psychomotor teaching levels of Education for SL Communication. The main contributions of the model to the teaching of SL are the promotion of the development of several CS on deaf children, allowing them to acquire knowledge through social interactions within a CMRLE and SLPMs (designed for a Rainbow Colors SL lesson at ASPAS). And finally, the model can be adapted to specific SL learning needs and can be reproduced as a complementary SLTM at other schools or deaf people Associations.

Mindful of psychomotor relationships between knowledge and communication, it is observed that there is a strong correlation coefficient of 0.99% (Fig.3B) between the SL repetitions from the one-hour Colors lesson and the number of Correct Answers given by the children. It is established that the more they practice SL positions (reading, writing and speaking through the SLPMs) the more they learn to communicate.

According to the Percentage of Development of SLCS and other CS (Fig.3A). The use of SLPMs has shown the following results: The use of videos has shown a 91.76% improvement in the SLCS, and the use of AR to develop speaking skills has shown an improvement of 95.60%. The use of the SL book has shown an improvement of 89.13% for reading skills and 86.36% for writing skills. This demonstrates that by using AR avatars, there is an increased level of interest in speech and makes the children feel more confident that they will learn to speak (Table 1, Q9). It also has been demonstrated that learning in collaboration with others increases learning achievement. (Table 2B).

With all the above, teachers have on their hands a complementary and adaptable SLTM which ensures the full understanding of concepts, meanings or ideas in accordance to different communication learning needs of deaf children.

## 5   Future Research

We consider to add a new SLPM based on Blender and OpenKinect camera for motion capture. A project at Microsoft Research China [10] has proved the recognition in real time of translating signs at the same time a person is performing them in front of the Kinect. By adding another SLPM to be used along with the MuCy model, the Teaching-Learning Process will be faster and efficient, because feedback to the learners is immediately streamed on the screens.

**Acknowledgments.** We would like to thank to ASPAS for all the support and to gave us the chance to offer an alternative SLTM to deaf children who will benefit from it.

## References

1. Adamo-Villani, N., Carpenter, E., Arns, L.: 3D Sign Language Mathematics in Immersive Environment. In: Proc. of ASM 2006 - 15th International Conference on Applied Simulation and Modeling, Rhodes, Greece, pp. 2006–2015 (2006)
2. Association of Parents of Deaf Children of Salamanca, ASPAS, Spain (2014), http://www.aspas-salamanca.es/ (viewed on July 6, 2013)
3. Blender. Blender 2.69 (2014), http://www.blender.org/ (viewed on September 1, 2013)
4. Billinghurst, M.: Augmented Reality in Education. New Horizons for Learning, Seatle WA, USA (2002)
5. Billinghurst, M., Kato, H., Poupyrev, I.: The Magic Book: A Transitional AR Interface. Computers and Graphics 25(5), 745–753 (2001)
6. Billinghurst, M., Kato, H., Poupyrev, I.: The MagicBook. Moving Seamlessly between Reality and Virtuality. Human Interface Technology Laboratory, University of Washington, Hiroshima City University and Sony C.S.Labs (2001)
7. Billinghurst, M., Kato, H., Poupyrev, I.: MagicBook: Transitioning between Reality and Virtuality. In: Proceeding of the Extended Abstracts on Human Factors in Computing Systems, New York, pp. 25–26 (2001)

8. Billinghurst, M., Kato, H., Poupyrev, I.: Collaboration with tangible Augmented Reality Interfaces. Human Interface Technology Laboratory, University of Washington, Hiroshima City University and Sony Computer Science Laboratories (2002)

9. Bloom, B.: Taxonomy of educational objectives. The classification of educational goals: cognitive domain. Handbook I. Longmans, Green, New York, Toronto (1956)

10. Chai, X., Li, G., Lin, Y., Xu, Z., Tang, Y., Chen, X.: Sign Language Recognition and Translation with Kinect. Key Lab of Intelligent Information Processing of Chinese Academy of Sciences (CAS), Institute of Computing Technology. Microsoft Research Asia, Beijing, China (2013)

11. Chaiklin, S.: Vygotsky's educational theory and practice in cultural context. The zone of proximal development in Vygotsky's analysis of learning and instruction. Cambridge University Press (2003)

12. Danielson, C.: The framework for teaching. Evaluation instrument. The Danielson Group (2013), http://www.danielsongroup.org (viewed on December 1, 2013)

13. Department of Education and Early Childhood Development. The Principles of Learning and Teaching P–12 Unpacked (2014), http://www.education.vic.gov.au (viewed on January 10, 2014)

14. Ivic, I.: Lev Semionovich Vygotsky. UNESCO 24(3-4), 773–799 (1994)

15. Mayberry, R.I.: Cognitive development in deaf children: the interface of language and perception in neuropsychology. In: Segalowitz, S.J., Rapin, I. (eds.) Handbook of Neuropsychology, 2nd edn., vol. 8, Part II (2002)

16. Mertzani, M.: Considering Sign Language and Deaf Culture in Computer Mediated Communication Environments: Initial Explorations and Concerns. In: 9th Theoretical Issues in Sign Language Research Conference, Florianopolis, Brazil (2008)

17. Milgram, P., Takemura, H., Utsumi, A., Kishino, F.: Augmented Reality: A class of displays on the reality-virtuality continuum. ATR Communication Systems Research Laboratoriesm, Kyoto, Japan. SPIE, vol. 2351 (1994)

18. Ó Dúill, M.: Teaching: Traditional versus Turing. In: X World Conference on Computers in Education, Torun, Poland (2013)

19. Unity Technologies, Unity3D V4.3 (2014), http://unity3d.com/unity (viewed on October 5, 2013)

20. Vuforia Developer. Vuforia^TM SDK, Unity extension Vuforia–2.8 (2014), https://developer.vuforia.com/resources/sdk/unity (viewed on December 22, 2013)

21. Vygotsky, L.: The principles of social education of deaf and dumb children in Russia. In: Proceedings of the International Conference on the Education of the Deaf, London, pp. 227–237 (1925)

22. Vuzix Corporation. Wrap 920AR and 1200DXAR Eyewear (2014), http://www.vuzix.com/ (viewed on December 22, 2013)

# Achievement Goals as Antecedents of Achievement Emotions: The 3 X 2 Achievement Goal Model as a Framework for Learning Environments Design

Margherita Brondino, Daniela Raccanello, and Margherita Pasini

Department of Philosophy, Education and Psychology, University of Verona (Italy),
Lungadige Porta Vittoria, 17, 37129 Verona, Italy
{daniela.raccanello,margherita.brondino,
margherita.pasini}@univr.it

**Abstract.** Achievement goals and achievement emotions, given their important role in learning processes, could affect the effectiveness of Technology Enhanced Learning (TEL) programs. In this study, the links between achievement goals and achievement emotions are explored. The participants were 466 Italian university students who completed a 48-item questionnaire about learning exam-relevant material. Confirmatory factor analyses showed the goodness of the hypothesized models for goals (task-approach, task-avoidance, self-approach, self-avoidance, other-approach, other-avoidance) and emotions (including enjoyment, relaxation, hope, pride, relief, anger, boredom, anxiety, shame, hopelessness). A path analyses indicated that task goals predicted activity- and outcome-related emotions, matched by valence; self-approach goals positively predicted one positive activity-related emotion; other-avoidance goals positively predicted activity- and outcome-related emotions. The results are discussed considering the utilities of the instruments for the design of TEL products, to assess motivation and affect involved in learning environments.

**Keywords:** Achievement Goals, Achievemente Emotions, Learning Environment Design, Emotions for TEL.

## 1    Introduction

Achievement emotions have recently roused particular interest within educational psychology for both their theoretical and applied importance, as constructs strictly linked to cognitive, motivational, and behavioural domains related to learning [8, 13]. Even if it could be difficult to observe emotional dynamics in technology-based learning environments, as happens in traditional learning contexts, also in those settings emotions have important influence on learning, engagement, and achievement. For instance, individual differences in achievement emotions affect students' learning choices, in terms of using online versus face-to-face learning modes [18].

T. Di Mascio et al. (eds.), *Methodologies and Intelligent Systems for Technology Enhanced Learning*, Advances in Intelligent Systems and Computing 292,
DOI: 10.1007/978-3-319-07698-0_7, © Springer International Publishing Switzerland 2014

According to the control-value theory [10], achievement emotions are focused on achievement activities or outcomes. Activity-related emotions pertain to on-going achievement related activities, while outcome emotions pertain to the outcome of these activities. The model posits that achievement emotions can be described in terms of both antecedents and consequences [10]. Among the antecedents, a key role is played by achievement goals, probably the most popular motivational construct investigated in the last two decades [6]. They can be defined as "cognitive-dynamic aims that focus on competence", comprising two dimensions: Definition–in terms of mastery and performance constructs–and valence–in terms of approach or avoidance strivings [4, p. 614]. Concerning relationships between the two constructs, the literature suggested associations between mastery goals and both activity-focused and outcome-focused emotions, and associations between performance goals and outcome-related emotions [11, 12, 14]. However, the hypothesized matching between valence of goals (as approach or avoidance strivings) and emotions (as positive or negative) was only partially supported.

Recently, a further distinction characterizing mastery goals and separating task-based and self-based goals has been proposed, resulting in a 3 X 2 model encompassing six goals: Task-approach goals, focused on acquiring competence related to the task; task-avoidance goals, focused on avoiding to demonstrate incompetence related to the task; self-approach goals, focused on acquiring competence based on the self; self-avoidance goals, focused on avoiding to demonstrate incompetence based on the self; other-approach goals, focused on acquiring competence comparing to others; and other-avoidance goals, focused on avoiding to demonstrate incompetence comparing to others [5]. However, scarce attention has been paid to the study of the relationships between these six goals and achievement emotions, including a wide range of emotions partially neglected by the literature, such as relief or relaxation.

### 1.1    The Present Study

Therefore, this work aimed both at exploring some psychometric properties of two instruments measuring achievement goals and achievement emotions in the Italian context, according to the more recent conceptualization of the two constructs [5, 10], and to extend current findings about their links. From an applied perspective, the development of valid questionnaires, characterized by clear language, brief administration, and simple answer format–such as multiple-choice questions, associated to low anxiety [13]–represents an opportunity also for Technology Enhanced Learning (TEL) environments, in which limitations related to time constraints or the need to assist individually people may pose methodological problems. Thus, in light of the pervasive use of technology within educational contexts, the two instruments at issue could be useful for the design of TEL products, in order to assess motivational and affective dimensions, before, during, or after learning.

This work is focused on two aims. The first aim was to investigate the psychometric properties of two instruments to assess achievement goals and achievement emotions: (a) A first Italian version of a questionnaire measuring achievement goals as conceptualized by the 3 X 2 model [5]; (b) An instrument comprising a large variety of achievement emotions, based on the control-value theory [10] and adapted from a previous work [15]. Both instruments referred to a specific setting in light of the documented contextual specificity of the two constructs [5, 9, 10, 17]. The second aim was to study the relationships between goals and emotions among university students. On the basis of literature [11, 12, 14], we expected task (approach/avoidance) goals and self (approach/avoidance) goals to predict both activity-related and outcome-related emotions, and other (approach/avoidance) goals to predict outcome-related emotions. We hypothesized matching between valence of goals and emotions.

## 2    Method

### 2.1    Participants

The participants were 466 university students attending the Faculty of Education at the University of Verona, Italy, coming from different social backgrounds (mean age: 23 years, range: 19-55 years; 93% female). They were tested in the autumn of 2011 (77%), in the spring of 2012 (2%), and in the autumn of 2012 (21%).

### 2.2    Material and Procedure

This work is part of a larger project, the "Minimal Knowledge Project", aiming at studying whether and how different cognitive, motivational, and emotional dimensions can predict future performance in university students attending the Faculty of Education at the University of Verona, Italy. The project has started in the autumn of 2011, and the data are currently gathered each year. The students are administered a questionnaire when taking a compulsory test during their first year. Passing the test, focused indeed on "minimal knowledge", is compulsory to enrol to their second year of course. In this paper, we present the data about 48 items measuring achievement goals and achievement emotions, referred to learning exam-relevant material. We also gathered data about studying, here not presented. The participants were asked to refer to the university subjects they were studying. They were assured about anonymity and participated on a voluntary basis. Each session lasted about fifteen minutes.

#### 2.2.1  Achievement Goals
They were assessed by means of an Italian adaptation of the 3 X 2 achievement goal questionnaire for taking an exam [5]. The instrument was translated from English to Italian and then from Italian to English according to the back-translation procedure, with the permission of A. J. Elliot. The final questionnaire included 36 items, to be evaluated on a 7-point Likert-type scale (1 = *not at all true for me* and 7 = *completely true for me*), about learning exam-relevant material, with three items referred to each of the six goals, namely task-approach goals (TAPG; e.g., *Learn many new concepts*),

task-avoidance goals (TAVG; e.g., *Avoid a superficial understanding of many topics*), self-approach goals (SAPG; e.g., *Learn better than I have learnt in the past on these types of tasks*), self-avoidance goals (SAVG; e.g., *Avoid learning worse than I normally learn on these types of tasks*), other-approach goals (OAPG; e.g., *Learn better than my classmates*), and other-avoidance goals (OAVG; e.g., *Avoid learning worse than other students*). The order of the items was the same of the original instrument.

### 2.2.2 Achievement Emotions

The students evaluated ten achievement emotions, conceptualized according to the control-value theory [10], on a 7-point Likert-type scale (1 = *strongly disagree* and 7 = *strongly agree*) [adaptation from 15]. For each emotion, three words were presented. The emotions included two positive activity-related emotions (enjoyment, relaxation), three positive outcome-related emotions (hope, pride, relief), two negative activity-related emotions (anger, boredom), and three negative outcome-related emotions (anxiety, shame, hopelessness). The order of the words was randomized and kept constant.

# 3      Results

## 3.1      Confirmatory Factor Analyses

We performed confirmatory factor analyses (CFA) to evaluate the psychometric properties of the two questionnaires, using AMOS 7.0 [2]. We took into account the following indexes: Chi-square, comparative fit index (CFI), incremental fit index (IFI), root-mean-square error of approximation (RMSEA), Akaike information criterion (AIC), and Bayesian information criterion (BIC). We considered the next threshold values: CFI and IFI $\geq$ .90, and RMSEA $\leq$ .08 [7].

For achievement goals, the results of the CFA indicated that the latent model based on the 3 X 2 distinction, in which items about the six goals load on six distinct latent factors, adequately fitted the data ($\chi^2(120) = 352.81$, CFI = .93, IFI = .93, RMSEA = .07, AIC = 454.81, BIC = 659.51). For achievement emotions, two CFA, separated by valence, enabled to test the goodness of two models in which items about the five emotions load on five distinct latent factors. The results indicated that the two latent models adequately fitted the data, separately for positive ($\chi^2(80) = 289.50$, CFI = .92, IFI = .92, RMSEA = .08, AIC = 369.50, BIC = 530.05) and negative ($\chi^2(78) = 289.59$, CFI = .92, IFI = .93, RMSEA = .08, AIC = 373.59, BIC = 542.17) emotions. On the whole, the goodness of fit indexes were acceptable, and all the loading factors were sufficiently high, i.e., higher than .54.

## 3.2      Internal Consistency, Descriptive Statistics, Intercorrelations, and Analyses of Variance

For both achievement goals and achievement emotions, all α-values were higher than .70 (except for relief), supporting homogeneity for each factor (Table 1). Students' responses on the items about the six goals and the ten emotions were averaged together (see Table 1 for descriptive statistics and intercorrelations).

Two repeated-measure analyses of variance (ANOVA), with type of goal or emotion as the within-subject factor, respectively, were run. The first ANOVA revealed a significant effect of goal, $F(5, 2040) = 409.86$, $p < .001$, $\eta p2 = .50$. Pair comparisons (with the Bonferroni correction) indicated higher values for task-approach goals compared to task-avoidance and self-approach goals (not differing), higher than self-avoidance goals, higher than other-avoidance goals, in turn higher than other-approach goals. The second ANOVA revealed a significant effect of emotion, $F(9, 3672) = 309.63$, $p < .001$, $\eta p2 = .43$. Pair comparisons indicated that, among positive emotions, enjoyment, hope, and pride had higher scores compared to relaxation and relief; among negative emotions, anxiety had the highest score.

## 3.3 Path Analysis

A path analysis was conducted using AMOS 7.0 to test the relationships between goals and emotions. The valence dimension for emotions was modelled applying the correlated uniqueness approach [3]. The results (Fig. 1) showed as task-approach goals predicted all the emotions but anxiety.

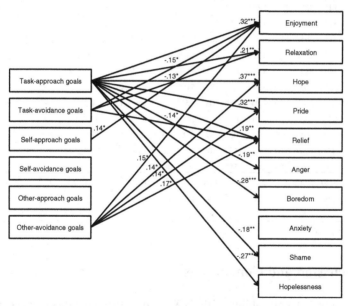

*Note. * $p < .05$, ** $p < .01$, *** $p < .001$.*

**Fig. 1.** The standardized paths of the hypothesized model

They positively predicted both positive activity-related emotions (enjoyment, relaxation) and positive outcome-related emotions (hope, pride, relief), while they negatively predicted both negative activity-related emotions (anger, boredom) and negative outcome-related emotions (shame, hopelessness). Task-avoidance goals negatively predicted positive activity-related emotions (enjoyment, relaxation) and a positive outcome-related emotion (relief). Self-approach goals predicted a positive activity-related emotion (enjoyment). Self-avoidance and other-approach goals did not predict any emotion. Finally, other-avoidance goals positively predicted both a positive activity-related emotion (enjoyment) and positive outcome-related emotions (hope, pride, relief). The emotions with higher level of accounted variability were hope (18.5%) and pride (18.2%), followed by enjoyment (12.4%) and hopelessness (9.2%), while the emotions with less explained accounted variability were relaxation (4.3%) and anxiety (0.9%).

**Table 1.** Intercorrelations, means $(M)$, standard deviations $(SD)$, and $\alpha$-values about achievement goals and achievement emotions

| | 1 | 2 | 3 | 4 | 5 | 6 | 7 | 8 | 9 | 10 | 11 | 12 | 13 | 14 | 15 | 16 |
|---|---|---|---|---|---|---|---|---|---|---|---|---|---|---|---|---|
| 1. TAPG | - | - | - | - | - | - | - | - | - | - | - | - | - | - | - | - |
| 2. TAVG | .63** | - | - | - | - | - | - | - | - | - | - | - | - | - | - | - |
| 3. SAPG | .51** | .58** | - | - | - | - | - | - | - | - | - | - | - | - | - | - |
| 4. SAVG | .35** | .55** | .61** | - | - | - | - | - | - | - | - | - | - | - | - | - |
| 5. OAPG | -.03 | .10* | .25** | .37** | - | - | - | - | - | - | - | - | - | - | - | - |
| 6. OAVG | .14** | .32** | .38** | .52** | .68** | - | - | - | - | - | - | - | - | - | - | - |
| 7. Enjoyment | .31** | .16** | .24** | .15** | .05 | .15** | - | - | - | - | - | - | - | - | - | - |
| 8. Relaxation | .17** | .05 | .13** | .07 | .06 | .05 | .54** | - | - | - | - | - | - | - | - | - |
| 9. Hope | .41** | .27** | .28** | .18** | .01 | .15** | .56** | .42** | - | - | - | - | - | - | - | - |
| 10. Pride | .39** | .32** | .27** | .16** | .08 | .19** | .63** | .37** | .64** | - | - | - | - | - | - | - |
| 11. Relief | .16** | .07 | .14** | .13** | .12* | .18** | .54** | .61** | .45** | .49** | - | - | - | - | - | - |
| 12. Anger | -.22** | -.14** | -.10 | -.03 | .14** | .08 | -.15** | -.12* | -.25** | -.15** | .14** | - | - | - | - | - |
| 13. Boredom | -.27** | -.14** | -.10* | -.01 | .11* | .04 | -.20** | -.04 | -.22** | -.18** | .11* | .68** | - | - | - | - |
| 14. Anxiety | -.01 | .01 | .05 | .06 | .08 | .07 | -.12* | -.43** | -.13* | -.03 | -.13** | .37** | .35** | - | - | - |
| 15. Shame | -.18** | -.10* | -.04 | .02 | .15** | .10* | .01 | -.08 | -.14** | -.07 | .17** | .54** | .45** | .42** | - | - |
| 16. Hopelessness | -.25** | -.14** | -.04 | .03 | .15** | .07 | -.14 | -.14** | -.28** | -.18** | .11* | .70** | .71** | .47** | .57** | - |
| $M$ | 5.49 | 4.90 | 4.84 | 4.07 | 2.85 | 3.48 | 4.08 | 3.34 | 4.58 | 4.37 | 3.24 | 2.13 | 2.30 | 3.72 | 2.19 | 2.42 |
| $SD$ | .98 | 1.12 | 1.28 | 1.35 | 1.32 | 1.51 | 1.21 | 1.24 | 1.18 | 1.20 | 1.11 | 1.08 | 1.04 | 1.37 | 1.22 | 1.13 |
| $\alpha$ | .77 | .70 | .80 | .71 | .83 | .81 | .81 | .81 | .77 | .76 | .68 | .77 | .70 | .75 | .81 | .79 |

*Note.* $* p < .05$, $** p < .01$.

## 4      Discussion and Conclusions

We examined the generalizability of the 3 X 2 model [5] to the Italian context and the validity of a questionnaire assessing achievement emotions [10], confirming the goodness of both instruments when used with university students about learning exam-relevant material. They will be proposed with an online methodology, as a TEL tool to monitor changes in motivation and emotions: During subsequent years of university, students will be administered longitudinally the same questionnaire, e-mailing an URL referred to an online version of the questionnaire, developed through the software ApsymSurvey [1]. Measures about achievement emotions could be integrated with graphical stimuli, particularly suitable for TEL environments [16].

We also extended current findings about the relationship between achievement goals and achievement emotions. First, **task** goals predicted both activity-related and outcome-related emotions, matched by valence (with some exceptions); self-approach goals positively predicted one activity-based emotion. These findings suggest an explanation for the partial inconsistency between the theoretical prediction of positive links between mastery goals and activity-related emotions, and the data about further links with outcome-related emotions [11, 12]. We partially agree with Putwain et al. [14], underlining the role of grades in determining emotions also whether the goal concerns mastery of competence. However, we argue that comparisons involving grades could be salient when the goal is focused on the task (with links to activity- and outcome-related emotions), while they could be less important for goals focused on the self (with links to activity-related emotions), about which individuals could take into account other indicators, not necessarily based on the outcome, but related to the on-going processes. Second, other-avoidance goals positively predicted positive activity- and outcome-related emotions, differently from the hypotheses. This could be due to the low salience of comparisons to others, as suggested by the low scores of other goals, for a frequently solitary activity such as learning exam-relevant material.

Considering the salience of both achievement goals and emotions in learners' daily life, our results might be useful from an applied perspective for the design of TEL programs. TEL environments, supporting a self-directed approach, could have a different impact on learners' performance, on the basis of different achievement goals or emotional orientation. Therefore, a preliminary assessment of such constructs could help to orientate individuals towards different learning pathways, adapt to their individual attitudes. However, future research should examine deeper this issue involving TEL contexts, given that evidence-based indications on how to design educational settings taking into account individuals' motivational and emotional dimensions are still scarce even for traditional learning contexts. In addition, a better comprehension of the links between achievement goals and achievement emotions is useful for the interpretation of the processes characterizing learning environments–resulting from complex interactions between cognitive, motivational, and affective dimensions–and consequently to a better TEL programs' design.

# References

1. Apsym Lab, ApsymSurvey (online survey software). Verona University, Italy (2013)
2. Arbuckle, J.L.: Amos (Version 7.0) (computer program). SPSS, Chicago (2006)
3. Byrne, B.M.: Structural equation modeling with Amos: Basic concepts, applications, and programming, 2nd edn. Taylor and Francis Group, New York (2010)
4. Elliot, A.J., Murayama, K.: On the measurement of achievement goals: Critique, illustration, and application. Journal of Educational Psychology 100, 613–628 (2008)
5. Elliot, A.J., Murayama, K., Pekrun, R.: A 3 X 2 achievement goal model. Journal of Educational Psychology 103, 632–648 (2011)

6. Graham, S., Weiner, B.: Motivation: Past, present, and future. In: Harris, K.R., Graham, S., Urdan, T., et al. (eds.) APA Educational Psychology Handbook. Individual differences and cultural and contextual factors, vol. 2, pp. 367–397. American Psychological Association, Washington, DC (2012)
7. Hair, J., Black, B., Babin, B., Anderson, R., Tatham, R.: Multivariate data analysis, 6th edn. Prentice-Hall, Upper Saddle River (2006)
8. Linnenbrink-Garcia, L., Pekrun, R.: Students' emotions and academic engagement: Introduction to the special issue. Contemporary Educational Psychology 36, 1–3 (2011)
9. Minnaert, A.: Goals are motivational researchers' best friend, but to what extent are achievement goals and achievement goal orientations also the best friend of educational outcomes? International Journal of Educational Research 61, 85–89 (2013)
10. Pekrun, R.: The control-value theory of achievement emotions: Assumptions, corollaries, and implications for educational research and practice. Educational Psychology Review 18, 315–341 (2006)
11. Pekrun, R., Elliot, A.J., Maier, M.A.: Achievement goals and discrete achievement emotions: A theoretical model and prospective test. Journal of Educational Psychology 98, 583–597 (2006)
12. Pekrun, R., Elliot, A.J., Maier, M.A.: Achievement goals and achievement emotions: Testing a model of their joint relations with academic performance. Journal of Educational Psychology 101, 115–135 (2009)
13. Pekrun, R., Stephens, E.J.: Academic emotions. In: Harris, K.R., Graham, S., Urdan, T., et al. (eds.) Individual Differences and Cultural and Contextual Factors. Individual differences and cultural and contextual factors, vol. 2, pp. 3–31. American Psychological Association, Washington, DC (2012)
14. Putwain, D.W., Sander, P., Larkin, D.: Using the 2 X 2 framework of achievement goals to predict achievement emotions and academic performance. Learning and Individual Differences 25, 80–84 (2013)
15. Raccanello, D.: Students' expectations about interviewees' and interviewers' achievement emotions in job selection interviews. Journal of Employment Counseling (in press)
16. Raccanello, D., Bianchetti, C.: Pictorial representations of achievement emotions: Preliminary data with primary school children and adults. Advances in Intelligent and Soft Computing (in press)
17. Raccanello, D., Brondino, M., De Bernardi, B.: Achievement emotions in elementary, middle, and high school: How do students feel about specific contexts in terms of settings and subject-domains? Scandinavian Journal of Psychology 54, 477–484 (2013)
18. Tempelaar, D.T., Niculescu, A., Rienties, B., Gijselaers, W.H., Giesbers, B.: How achievement emotions impact students' decisions for online learning, and what precedes those emotions. The Internet and Higher Education 15(3), 161–169 (2012)

# The Relationship between Psycho-pedagogical and Usability Data in the TERENCE Evaluation Methodology

M.R. Cecilia[1], V. Cofini[1], Tania di Mascio[2], and Pierpaolo Vittorini[1]

[1] Dep. of Life, Health and Environmental Sciences, Univ. of L'Aquila, Italy
[2] Dep. of Information Engineering, Computer Science and Mathematics,
Univ. of L'Aquila, Italy
mariarosita.cecilia@graduate.univaq.it,
{tania.dimascio,pierpaolo.vittorini}@univaq.it,
vincenza.cofini@cc.univaq.it

**Abstract.** The TERENCE project was an European FP7 project that improved reading comprehension of its primary users, namely, 7–11 years old children, hearing and deaf[1], from primary schools, with deep text comprehension problems. The project developed an adaptive learning system (ALS) delivering learning material and tasks in an adaptive fashion, according to the profiles of its learners. The paper describes the TERENCE evaluation methodology, its innovative aspect (i.e., the parallel investigation of psycho-pedagogical effectiveness and usability), and focuses on the relationship between usability and psycho-pedagogical data. The study shows that precision in smart games is linearly related to comprehension.

**Keywords:** effectiveness, usability, evaluation.

## 1 Introduction

Developing the capabilities of children to comprehend written texts is key to their development as young adults. From the age of 7-8 until the age of 11, children develop as independent readers (novice compreheneders). Nowadays, more and more children in that age range turn out to be poor comprehenders, i.e., they have well developed low-level cognitive skills (e.g. vocabulary knowledge [1,2]), though they have difficulties in deep text comprehension (e.g. inference making [3]). "Poor comprehenders" are estimated to be 5% to 10% of novice comprehenders [4]. The estimate dramatically increases when the whole population of hearing-impaired children is considered. For instance [5], estimates that only 19% out of 504 hearing impaired 7-20 old have reading comprehension scores above the third grade level. The consequences of unremediated reading comprehension difficulties extend beyond literacy skills, and can have a negative impact, for instance, on motivations to reading, performances in science curricula and a child's self-esteem [6]. However finding

---

[1] Improvement in reading comprehension for deaf learners was statistically significant only in the Italian case.

T. Di Mascio et al. (eds.), *Methodologies and Intelligent Systems for Technology Enhanced Learning*, Advances in Intelligent Systems and Computing 292,
DOI: 10.1007/978-3-319-07698-0_8, © Springer International Publishing Switzerland 2014

stories and educational material that are appropriate for poor comprehenders is a challenge and educators are left alone in their daily interaction with poor comprehenders. Most educational material for novice comprehenders is paper based, and is not easily customisable to the specific needs of poor comprehenders. A few adaptive learning systems (ALSs) promote reading interventions, but they have high-school or university textbooks as reading material, instead of stories, and are developed for old children or adults, and not specifically for poor comprehenders, hearing and deaf.

TERENCE aimed at filling such a gap. It was a European FP7 ICT multidisciplinary project and developed an adaptive learning system (ALS) that delivers learning material and tasks in an adaptive fashion, according to the profiles of its learners. All together, they constitute the so-called TERENCE world. The TERENCE world design is based on its users' requirements and revised in light of evaluation sessions with learners, educators and domain experts, in an interactive manner, as the UCD suggest. In particular, its learning material contains books of stories, rewritten into 4 levels of difficulties for all the TERENCE learners, and games that serve both to reason about the story mainly in a visuo-perceptual manner (called smart games), and to take a "break" from reading and still stimulating visuo-perceptual skills necessary for interacting with the TERENCE world (called relaxing games).

Several evaluation sessions took place. All of them took into account the need for developing a usable and effective system, and gave valuable feedback for revising the software and improving the knowledge of poor comprehenders. It is worth remarking that TERENCE – as a result of the final evaluation session – actually improved reading comprehension of 7–11 years old children, hearing and deaf1   [7].

The paper describes the TERENCE evaluation methodology, shows its innovative aspect, i.e., the combined investigation of usability and psycho-pedagogical effectiveness, and focuses on the results of concurrently analysing the relationship between usability and psycho-pedagogical data. Usability is the effectiveness, efficiency and satisfaction with which specified users achieve specified goals in particular environments.

## 2      The TERENCE Evaluation Methodology

The innovative aspect of TERENCE is its evaluation methodology. Evaluating TERENCE consisted in assessing both the psycho-pedagogical effectiveness and the usability of the overall TERENCE world. Its characteristic lies in a combined investigation of the two aspects.

The evaluation in TERENCE was articulated in two expert-based evaluations and two user-based evaluations, in an iterative manner, as the user centered design methodology suggests.

The first evaluation, run from April to June 2012, was carried out by domain experts of (poor) text comprehension and human computer interaction (1st expert-based evaluation) and evaluated the initial learning material and the preliminary prototypes. The second evaluation (small-scale evaluation) was run during the summer 2012, and focused on the 1st software prototype. The third evaluation was performed between November 2012 and January 2013, where several experts of the Consortium revised the new learning material as well as the new prototypes (2nd expert-based evaluation). The fourth (and final) evaluation (large-scale evaluation), run from March to June 2013, was carried out with end-users and finally investigated the two aspects of the TERENCE world: its pedagogical effectiveness, its usability and, more specifically, the user experience.

The research hypothesis of the large-scale evaluation was that the TERENCE world significantly improves the text comprehension of the TERENCE main learners, hearing or deaf. The large-scale evaluation was designed and developed with the direct involvement of TERENCE learners at school. In all situations, data were gathered during the usage of the TERENCE system at school. The term "large" refers to the huge number of participating learners (hearing and deaf), c. 800, across Italy and UK. For gathering data during the usage of the system, the consortium developed a common protocol in collaboration with the educational bodies and involved teachers of both countries, to be hosted in the school structures. The inclusion criteria was: primary school children, male and female, within 7-11 years old. The exclusion criteria was the lack of informed consent. The protocol was organised into three main phases:

- **Phase 1: PRE Evaluation Phase** - Participants were assessed via psychological tests so as to evaluate the initial comprehension level of the TERENCE learners and to properly initialize the TERENCE system, in particular its adaptive engine. This phase was related only to the pedagogical part of the study;
- **Phase 2: Stimulation Phase** - Participants used the TERENCE software in terms of two sessions per week. Each session consisted in reading a story, then playing with the correlated smart games, finally playing with the relaxing games. Each child used the TERENCE system via his/her device. Each session was run collectively in a dedicated classroom. In Italy, sessions took place under the vigilance of the class teachers and the TERENCE tutors. The number of tutors varied according to the number of groups participating in the intervention on a daily basis. In UK, given budget constraints, only teachers controlled the learners. Accordingly, a higher compliance to the intervention plan was expected in Italy. This phase was related to both the pedagogical and the usability part of the study;
- **Phase 3: POST Evaluation Phase** - At the end, all children were re-tested like in Phase 1, so to evaluate the final comprehension level. This phase was related only to the pedagogical part of the study.

Therefore, in summary, the large scale evaluation was performed through a pre-post, controlled, non-randomized study. In Italy, the sample of hearing participants was made up of two schools, "Avezzano" and "Pescina". In the paper we refer only to the school of Pescina (Istituto Comprensivo "Fontamara"), because was the only one, among all recruited, who gave the authorisation to combine the usability data with the psycho-pedagogical data. All data were always gathered anonymously, in full compliance with the Italian Privacy Laws. Given the aforementioned evaluation methodology, we now describe the results that come out from the merge of usability and effectiveness data in the Istituto Comprensivo "Fontamara", Pescina, Italy.

## 3     The Relationship between Psycho-pedagogical and Usability Data

This section describes the relationship between usability and psycho-pedagogical data collected during the large-scale evaluation. Table 1 summarizes the variables investigated in the section, divided into usability and effectiveness variables.

The precision indicates the number of the games correctly resolved divided by the number of games played by a learner. The average reading time (AVG_RT) indicates the total amount of reading time in seconds divided by the number of the episodes read. The average playing time (AVG_PT) indicates the total amount of time in seconds spent in smart games divided by the total number of the smart games played by a learner.

**Table 1.** The variables investigated in the study

| Usability | Effectiveness |
|---|---|
| The number of stories read (SR) | The comprehension at the beginning of the intervention (CB) |
| The number of episodes read (ER) | The comprehension at the end of the intervention (CE) |
| The average reading time (AVG_RT) | Whether a child is a poor comprehender (PCB) or not (GCB)at the beginning of the intervention |
| The number of smart games played (SG) | Whether a child is a poor comprehender (PCE)or not (GCE)at the end of the intervention |
| The average playing time (AVG_PT) | |
| The difficulty level (1, 2 3, or 4) of the story assigned by the system at the end of intervention (RC_END) | |
| The precision in the smart games (pSG) | |

Reading comprehension (for variables CB and CE) was measured with the Italian "MT" standardized test [8]. The MT tasks are a measure of the reading and comprehension ability in scholar age. MT tasks were submitted to the sample by six clinical psychologists unaware about the objectives of the intervention. All group underwent to pre- and post- test evaluation. Each subject was evaluated in 40 minutes (both reading and comprehension trials). The cognitive evaluation sessions were conducted in dedicated rooms of educational institutions. One month was needed in order to complete each phase of the evaluation. The pre-test was carried out in January/February 2013 and the post-test in May/June 2013. The variables related to reading and not to comprehension are not investigated in this study. In detail, comprehension is assessed by letting the child read a story and answer a set of questions. Depending on the answers, a score (usually) ranging from 0 to 10 is assigned. The higher the score, the better the comprehension. Depending on the class and on the score, a child can be assigned to one of the following clusters: "Need for immediate intervention", "Attention is needed", "Sufficient performance", "Complete performance". A poor comprehender is a child belonging to one of the first two clusters (for variables PCB and PCE). The statistical analyses reported below were performed through the software Stata12/ME. To compare CB and CE, since both variables are ordinal, non-normal, and paired, a Wilcoxon signed rank test was used. Furthermore, for assessing whether usability data may predict psycho-pedagogical results, a multivariate linear regression was adopted [9].

The sample is made up of 68 hearing participants in scholar age, for whom there are both MT score and TERENCE scores. The sample includes 29 females (42.65%) and 39 males (57.35%), aged between 80 and 129 months (104± 13.78). Furthermore, 17 students belonged to the second class (25%), 16 students to the third class (23.53%), 20 students to the fourth (29.41%), and 15 students to the fifth class (22.06%) (Table 2).

**Table 2.** Socio-demographic characteristics of the sample

| Socio-demographic characteristics | Total sample (N=68) | |
|---|---|---|
| **Gender** | **n** | **%** |
| Male | 39 | 57.35 |
| Female | 29 | 42.65 |
| **Age** | **Mean (in months)** | **SD** |
| | 104 | 13.78 |
| **Class** | **n** | **%** |
| II | 17 | 25.00 |
| III | 16 | 23.53 |
| IV | 20 | 29.41 |
| V | 15 | 22.06 |

The analysis of the scores obtained from the "MT" standardized test for comprehension shows that 14 students (20.59%) resulted poor-comprehenders at the beginning of the intervention. They are equally distributed between the two sexes (7 females and 7 males; $\chi^2$=0.3897, p=0.532). Only 6 learners (8.82%) were poor comprehenders at the end of the intervention, 2 females and 4 males (Figure 2). There is no relationship between gender and poor comprehension ($\chi^2$=0.2334, p=0.629). A Wilcoxon signed rank test asserts that the difference in comprehension is statistically significant (z=-4.904, p<0.0001). Further results about the effectiveness of the intervention can be found in [7]. The participants read circa 14 stories and circa 84 episodes. The average reading time was less than a minute per page. The minimum value (7.91 seconds) suggests that some children did not read the episode. The users also resolved circa 131 smart games, circa thirty seconds per smart game. The average difficulty level of the story assigned by the system at the end of intervention was high. The precision in the smart games ranged from 0.55 to 0.96. Table 3 summarizes the psycho-pedagogical and the usability characteristics of the sample.

**Table 3.** Characteristics of the sample for the psycho-pedagogical variables and for usability variables

| | Total sample (N=68) | | | | | |
|---|---|---|---|---|---|---|
| | **Variable** | **Obs** | **Mean** | **SD** | **Min** | **Max** |
| | SR | 68 | 13.52941 | 2.249 | 6 | 21 |
| | ER | 68 | 84.48529 | 14.436 | 38 | 130 |
| Usability | AVG_RT | 68 | 47.65643 | 20.610 | 7.91 | 102.02 |
| | SG | 68 | 130.9706 | 20.359 | 63 | 180 |
| | AVG_PT | 68 | 30.56647 | 5.8435 | 20.85 | 49.56 |
| | RC_END | 68 | 3.205882 | .955 | 1 | 4 |
| | pSG | 68 | .7915208 | .095 | .55 | .96 |
| Psycho-pedagogical | CB | 68 | 7.617647 | 1.693 | 4 | 10 |
| | CE | 68 | 8.264706 | 1.192 | 5 | 10 |
| | **Variable** | **Obs** | **%** | | | |
| | GCB | 54 | 79.41 | | | |
| | PCB | 14 | 20.59 | | | |
| | GCE | 62 | 91.18 | | | |
| | PCE | 6 | 8.82 | | | |

Multivariate linear regression was used to examine whether usability data may predict psycho-pedagogical results. The analysis indicates that the comprehension at the end of the intervention (CE) was not associated with the stories read (SR), the episodes read (ER), the average reading time (AVG_RT), the difficulty level of the story assigned to each child by the system according to his/her level of comprehension at the end of intervention (RC_END), the smart games resolved (SG) and the average playing time (AVG_PT). On the other hand, there is a statistically significant association between the comprehension at the end of the intervention (CE) and the measure of the precision in the smart games (pSG) (t=2.49, pr=0.015). See Table 4 for details, and Figure 1 for the related regression graph.

**Table 4.** The relationship between the comprehension at the end of the intervention and usability data ($R^2$=0.1773)

| *Independent variable* | **CE** | **t** | **p** |
|---|---|---|---|
| | SR | -1.31 | 0.194 |
| | ER | 0.88 | 0.381 |
| | AVG_RT | -1.01 | 0.317 |
| *Dependent variables* | RC_END | 0.22 | 0.829 |
| | SG | 1.13 | 0.261 |
| | AVG_PT | 1.56 | 0.124 |
| | **pSG** | **2.49** | **0.015** |

## 4    Conclusions

TERENCE was an European project that improved reading comprehension of 7–11 years old children. In the paper, we initially described the TERENCE evaluation methodology and its innovative aspect of combining psycho-pedagogical effectiveness and usability investigation. Then, we reported on the relationship between usability and psycho-pedagogical data, by showing that precision in smart games is linearly related to comprehension. Even though a statistically significant relationship between CE and pSG was found, we cannot state that the prediction is reliable, since the $R^2$ value is quite low. This result may be explained by the fact that the sample may not be large enough. Another possible explanation is that – as our psychologists noticed – some children with comprehension problems read more than once the same story, learned the solutions of the smart games, replied them in the successive interactions with TERENCE, without actually having comprehended the story, only to quickly move to the relaxing games. For them, therefore, we may have experienced high precision values, without an actual improvement in comprehension. Furthermore, it is also reasonable that – simply – the precision is not sensible enough and/or specific to diagnose comprehension problems.

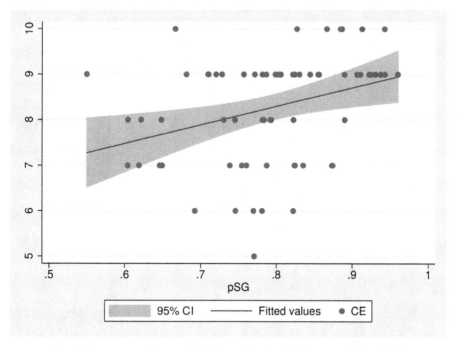

**Fig. 1.** Linear regression graph of comprehension (CE) wrt precision (pSG)

# References

1. Cain, K., Oakhill, J.V., Barnes, M.A., Bryant, P.E.: Comprehension Skill, Inference Making Ability and their Relation to Knowledge. Memory and Cognition 29, 850–859 (2001)
2. Marschark, M., Sapere, P., Convertino, C., Mayer, C., Wauters, L., Sarchet, T.: Are Deaf Students' Reading Challenges Really About Reading? American Annals of the Deaf (2009)
3. Yuill, N., Oakhill, J.: Effects of inference awareness training on poor reading comprehension. Applied Cognitive Psychology 233 (1988)
4. Lyon, G.R., Fletcher, J.M., Barnes, M.C.: Learning Disabilities. In: Mash, E.J., Barkley, R.A. (eds.) Child Psychopathology. The Guilford Press, NY (2003)
5. Wauters, L.N., van Bon, W., Tellings, A.E.J.: Reading Comprehension of Dutch Deaf Children. Reading and Writing 19, 49–76 (2006)
6. Cain, K.: Making sense of text: skills that support text comprehension and its development. Perspectives on Language and Literacy 35, 11–14 (2009)
7. Cofini, V., Di Giacomo, D., Di Mascio, T., Gennari, R., Vittorini, P.: The Pedagogical Evaluation of TERENCE: Preliminary Results for Hearing Learners in Italy. In: Proceedings of ebuTEL 2013, Trento (2013)
8. Cornoldi, C., De Beni, R., Pazzaglia, F.: Reading comprehension difficulties: Processes and interventions. In: Profiles of Reading Comprehension Difficulties: An Analysis of Single Cases, pp. 113–136. Lawrence Erlbaum Associates (1996)
9. Riffenburgh, H.R.: Statistics in Medicine, 3rd edn. Academic Press (2012)

# Many Children and Short Project Timing: How TERENCE Harmonized These Conflicting Requirements

Tania di Mascio[1], Alessandra Melonio[2], Laura Tarantino[1], and Pierpaolo Vittorini[3]

[1] DISIM, University of l'Aquila, V.le Gronchi, 18, 67100 L'Aquila, IT
{tania.dimascio,laura.tarantino}@univaq.it
[2] CS Faculty, Free University of Bozen-Bolzano, P.za Domenicani, 3, 39100 Bolzano, IT
alessandra.melonio@unibz.it
[3] MeSVA, University of L'Aquila, Viale S. Salvatore, Ed. Delta 6, 67100 Coppito, L'Aquila, IT
pierpaolo.vittorini@univaq.it

**Abstract.** Involving children in the software design and development processes (using e.g., the user centered and the participatory design methodologies) is nowadays considered a key factor to obtain an accessible and usable technology enhanced learning system; but how children have to be efficiently involved is still an open question. In this paper we report the experience of a FP7-ICT European project, TERENCE, aimed at developing a system for improving text comprehension in children 7-11 years old. Our experience suggests to extend the repertoire of inquiry techniques with a new one able to harmonize two conflicting requirements of real projects: many children to involve vs project strict timing.

## 1 Introduction

About 10% of young children are estimated to be poor text comprehenders: they are proficient in word decoding and other low-level cognitive skills, but show problems in deep text comprehension. The comprehension process may be stimulated by educational interventions carried out by primary school educators, who, among others, aim at retracing temporal and casual-temporal relations among main events of a story. In this respect, experiments show the pedagogical effectiveness of inference-making questions centered on a number of identified skills, together with adequate visual aids. Unfortunately, finding stories and educational material appropriate for poor comprehenders is a challenge and the few systems promoting reading interventions are based on high school or university textbooks. The TERENCE project, a European FP7–ICT project, aimed at filling this gap by developing and evaluating the first ad hoc Adaptive Learning System (ALS), in Italian and in English, for improving the reading comprehension of 7–11 years old poor comprehenders, who are the primary users of the system, building upon effective paper-and pencil reading strategies, and framing them into a playful and stimulating pedagogy-driven environment [1]. Secondary users of the system are educators (parents and teachers) and experts that design learning material made of stories and games: smart games are used for stimulating inference-making about stories, and relaxing games are used for motivating learners, according to a stimulation pedagogical

T. Di Mascio et al. (eds.), *Methodologies and Intelligent Systems for Technology Enhanced Learning*, Advances in Intelligent Systems and Computing 292,
DOI: 10.1007/978-3-319-07698-0_9, © Springer International Publishing Switzerland 2014

plan identified by experts. To achieve its goals, the project had to be multidisciplinary in nature, requiring input from many diverse area of expertise, from learning theories to software engineering, from evidence-based theories to Human Computer Interaction (HCI), as typical of Technology Enhanced Learning (TEL) system [9]. The project is just finished and we are able to draw conclusions on how choices about methodological issues impacted on the achieved results and on the success of the project. In particular, in this paper we want to analyze our experience with respect to the involvement of children in data gathering activities.

**Involving Children in the Design.** Technology is successful when it is well contextualized and based on a clear understanding of needs and behaviors of the persons using it. It is desirable that the design is conducted according to methodologies that actively involve users in the design process, like User Centered Design (UCD) or participatory design [9], which have also the positive side effect of favoring the integration of designed artifacts into organizations new to digital innovations, as in the case of schools. But, if it is reasonable to expect that contextual studies are conducted with users to elicit insights on their demands and tasks, it is legitimate to ask whether existing methods for the analysis of the context of use are really adequate for interacting with children. Actually, as expert recommends, data gathering methods cannot be used with children "as is": e.g., children might become anxious at the thought of taking a test and tests may conjure up thoughts of school [5]. Based upon an analysis of the literature and her own research experience with children, Druin [2] singled out four mail roles that children can play in the technology design process: user, tester, informant, and design partner. Conforming to this model, methods have been proposed according to co-design or participatory design approaches [4,5,13]. Unfortunately, the most important case studies regarding the involvement of children in technology design are lab-based research-oriented experiments conducted with no strict time and budget limits and no strict constraints related to (i) regular school activity timing, (ii) real project timing (as in FP7 projects) and (iii) psycho-pedagogical constraints. For example, in [3] it is explicitly said that the reported research was focused on what happens with children and technology outside of the school environment, because schools are generally "adult controlled" places. On the other hand, it has to be observed that in the case of children-oriented TEL systems, it is mandatory to ensure pedagogical effectiveness through large-scale studies and, in a real project with time and budget limitation, this requirement naturally implies school involvement. Furthermore, it may be the case that a TEL system is conceived to be integrated within regular school activities. Indeed, there are examples of co-design at school (e.g., in [13] authors explore the application of co-design methods with children 7-9 aged), but when situated within school activities co-design has some limitations if it is done with many learners and ethical and organizational constraints: for example, in TERENCE, schools imposed that all children of a class had to be involved at the same time and that the timing of data gathering activities had to be below one hour.

**The TERENCE Approach.** For the analysis of the context of use, in TERENCE we were hence forced to conceive innovative approaches able to harmonize conflicting requirements, among which: many children vs strict timing, controlled school environment vs sense of freedom. It has to be observed that, while adults have perceived computers for decades as workplaces, children are used to interact with technology embedded into toys [11] and in general within playful activities. Furthermore, theoretical and empirical studies show that learners are more motivated to participate in school-class activities if they are shaped like games (e.g., [6,14]). In [10] authors overview research findings about the correlations between the appeal of games and the psychological need satisfaction they provide, and propose a motivational model that shows that, besides the basic elements of *move* of the player and *outcomes* showing progresses, at least three factors determine engagement: *autonomy*, amounting to a sense of choice and psychological freedom, *competence*, realized by carefully balancing the game challenges to the players' skill, and *relatedness needs*, attained by stimulating collaboration or competition. All this considered, we decided to base on games not only the ALS stimulation plan but also the field study, and designed and experimented an innovative children-oriented data gathering approach, named *game-based user investigation*, based on game-full activities designed according to state-of-the-art motivational models [10]. The investigation involves both a direct observation by investigator participating to game administration and an indirect observation on collected material (e.g., game results, notes, audio/video recording). In the reminder of this paper we report, in Section 2, our experience of TERENCE field studies; in Section 3 we formalize such experience in a structured manner; finally, in Section 4, we compare the proposed method with traditional data gathering techniques, highlighting pro and cons, and discussing how its adoption contributed to the success of the system.

## 2 The TERENCE Experience

To guide the design and development of the TERENCE system, we adopted the user-centred design (UCD) methodology, which involves the end-users into the project from the very beginning, and aims at the overall usability of the system. The overall design process proceeded iteratively, by applying the following steps for each of the four generations of prototypes/system: (1) analysis of Context of Use and Users' Requirements (CUUR); (2) design of learning material, tasks and GUI prototypes, and (3) evaluation.

CUUR was conducted through a preparatory study followed by two rounds of field studies [12,7,8], in the UK and Italy (see for details Table 1). For the 2nd round some kind of direct interaction with learners was crucial to gather high quality data; furthermore to ensure pedagogical effectiveness of the system, a large-scale study was mandatory. The studies hence involved 2 schools in UK and 5 in Italy (learners aged 7-11) and were run as part of regular school activities. This resulted in a total of 20 primary school classes with children of all ages. In each class, we spent about 60 minutes with learners, and each class teacher participated in a 20-minutes interview.

**Planning Stage.** Before conducting experiments, games have been designed according to a methodology illustrated in Section 3 and the protocol of the game session, composed of different game-based activities, was checked and assessed with schoolteachers (e.g., if a challenge was deemed too difficult or too boring for a school class, it was revised according to teachers' feedback); we also gained the schoolteachers cooperation; we collected the informant consensus by legal representatives (e.g., parents, tutors) of the involved children; and we finally established time and places for the data gathering. We designed 6 games activities: 2 collaborative games, involving all class learners at

**Table 1.** An overview of TERENCE CUUR studies. Ex. Ed, L, and P indicate the number of experts, educators, learners and parents respectively.

| Study | Ex | Ed | L | P | Methods | Goals | Period |
|---|---|---|---|---|---|---|---|
| Preparatory Study | 5 | - | - | - | Interview, Brainstorming | How children are assessed by psychologists as poor comprehenders | May 2010 – Nov. 2010 |
| 1st Round of Field Study | - | 68 | 96 | 13 | Diary, Contextual inquiry | Constraints, main requirements of learning material, and first characterization of learners | Oct. 2010 – Jan. 2011 |
| 2nd Round of Field Study | - | 20 | 532 | - | Game-based user investigation | Redefinition of system user's types into classes of users, defining associated personas, according to requirements relevant for the adaptive engine of the ALS | Feb. 2011 – June 2011 |

the same time, and 4 single-player games. For each game activity, we created (1) the game materials (e.g., for the "Characters" game we prepared cards illustrating the main video game characters like Mario Bros), (2) the notes templates, and (3) the database to be populated by data we gathered. Each game activity has been designed with different cognitive load (warm-up, peak, and relaxing); see Table 2, reporting name, type, investigation topics and cognitive load of all administered games.

**Table 2.** The administered games

| Name | Type | Investigation Topic: Gather information about... | Cognitive load |
|---|---|---|---|
| Extracurricular activities | single-player | ... learners' favorite extracurricular activities. | warm-up |
| Technology use | single-player | ... learner use of TV, computer, and mobile phone. | peak |
| Characters | single-player | ... learners' favorite game characters. | relaxing |
| Interaction with parents | single-player | ... interaction between learners and their parents. | warm-up |
| Homework | collaborative | ... learners' homework. | relaxing |
| Console | collaborative | ... learners' favorite game console. | peak |

Autonomy, competence and relatedness needs were pursued across the various games. Autonomy was elicited by allowing learners to choose among several options for tackling a challenge or to take the decision to skip it. Competence was pursued by stimulating diverse skills across games (e.g., some games required mainly verbal skills whereas others mainly drawing skills).All games are compliant with playfulness, child personal enrichment, and ethical issues; each specific game includes a rewarding mechanism (e.g., marshmallows); and individual games should produce children-generated collectable results (e.g., conceptual maps).

A framework was created specifying the goals and moves of the game, and how autonomy, competence and relatedness needs are pursued. In Figure 1 we provide an example of instantiation of the framework for the "Characters" game associated to a specific topic to be investigated.

| Investigation topic | Gather information about the learners' favorite game characters (useful for designing avatars of TERENCE stimulation plan games). | CARDS |
|---|---|---|
| Game description | **Goal.** The goal of the challenge is to describe popular video game characters.<br>**Moves.** Each learner has to choose a card from the container. A card depicts a character of a popular console game. The entire class then discusses what they like or dislike about that character. | |
| Autonomy | Each learner can choose whether to extract the card and participate, or not, in the game; each learner can choose what to tell about the selected character. | |
| Competence | Each learner can express his own verbal skills. | |
| Relatedness needs | Each learner can feel part of the class by talking about characters or listening to others' preferences. | |

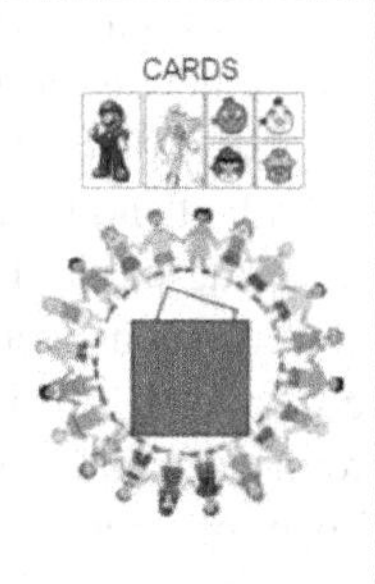

**Fig. 1.** Example of instantiation of the framework for the "Characters" game

**Running Stage.** During the TERENCE data gathering, two investigators observed 20 entire classes and then they performed 20 game sessions. During one game session, they administered the 6 games shown in Table 2 according to a flexible plan that takes into account: estimated duration of the games, remaining time, topic coverage, topic priorities, warm-up/peak/relaxing cognitive curve, number of involved children.

First of all, in each session, investigators introduced themselves, explained the aim of the session, and established a playful atmosphere (*nurturing phase*). It was made clear that the participation to games is free. Furthermore, investigator explained to children about the importance of the game session to ensure a sense of responsibility, essential to get reliable data (*motivational phase*). After that, the main *body phase* of the observation began: investigators administered games and observed children. For each administered game goals, moves, and rewards were introduced and excitement was provoked (*energizing step*). Then, the main direct observation took place: the specific game were administered with investigators keeping focus on how children carried on game activities, while stimulating children in maintaining interest and supporting their requests. Investigators tried to be aware of influences affecting children, took notes of each behavior interesting for later analysis, and took photos of the game areas (*playing step*). At the end of the specific game, investigators officially close the game, declare winners for group games when planned, and delivered prizes (e.g., marshmallows) (*rewarding step*). Afterwords, investigators collected and organized the game produced material (*reorganizing step*).

For ethical and motivational reasons, at the end of the game session (*closing phase*) we made sure that each child got a reward. Finally, investigators reordered collected material and wrote down first impressions about the experience before the analysis. Investigators also spent some time with schoolteachers that attended the game sessions to clarify and solve any doubt.

**Reporting Stage.** Since the method produced vast amount of high quality data, investigators used collected data to populate the data-base designed in the planning stage, and conducted statistical analysis to produce user classification, personas design, and requirements specification.

## 3    Game-Based User Investigation Technique

In this section we formalize in a structural manner the experience so far reported, in order to describe a new data gathering method able to harmonize conflicting require-ments arising in real projects which need the involvement of a high number of children while fulfilling budget and time limits. As in more traditional data gathering techniques the application of the method requires a sequence of three stages: planning, running, and reporting. A crucial aspect faced during the *planning stage* is the design/realization of the so called "investigator kit", based on the outcome of a preliminary study neces-sary to acquire in-depth knowledge on topics/subtopics to be investigated, appropriate language and way of approaching, context constraints. The kit includes games, game materials, customary notes templates, and a database to be populated by data gathered during the running stage. Depending on the established schedule, the running stage may consist of a number of independent game sessions (see Figure 2), each based on the same "investigator kit". The investigator kit is designed according to the following requirements: (i) there must be a specific game for each specific topic to be investigated; (ii) the overall set of games has to include games with different cognitive load so that a game session can mirror customary warm-up, peak, and relaxing interview phases; (iii) topics (and associated games) have to be prioritized according to their relevance to the project in order to be able to shape the individual game session on the fly at run-ning time while maintaining the warm-up/peak/relaxing structure (necessary to adjust the session depending on interrupts and other unpredictable events); (iv) the estimated duration of a game session should not exceed 60 minutes.

Design of individual games has to consider a number of factors: (a) each specific game must cover all subtopics of the topic it is associated to; these subtopics are the primary inspiration for the creation of the game that, in any case, has to be shaped according to consolidated game frameworks; (b) mandatory characteristics of individual games are: playfulness, child personal enrichment, compliance with ethical issues; (c)

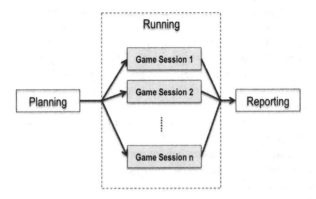

**Fig. 2.** Overall structure of data gathering activities, with focus on the running stage

each specific game must include a rewarding mechanism, designed so to stimulate the production of genuine data from each child; and (d) individual games should produce children-generated collectable results (e.g., conceptual maps).

As depicted in Figure 2, the *running stage* may consist of a number of independent game sessions based on the same "investigator kit". Each session includes the four phases of *nurturing, motivation, body, and closing*, according to the structure depicted in Figure 3. In particular, as to the body of a single game session, each administered game requires the four steps of *energizing, playing, rewarding and reorganizing*, giving rise to the overall iterative structure of the body phase in Figure 3. The experience described in the previous section provides example instantiations of phases and steps.

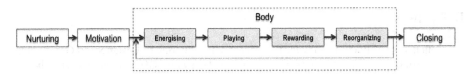

**Fig. 3.** The structure of one session, composed of four phases: nurturing, motivation, body, and closing, whit focus on the body phase

## 4   Conclusions

In this paper we discussed our experience within the TERENCE project, which is developing an ALS for supporting poor comprehenders and their educators, focusing on data gathering issues. The age of learners, along with literature studies on children involvement in school activities and in co-design, suggested not to use traditional methods indicated us to explore a game-based approach as primary data gathering method. This method embodies characteristics from other traditional investigation techniques: as questionnaires it allows to gather a high quantity of data in relatively short sessions, as interviews it allows a direct interaction with users, as user observation/field study it allows to view users in their real context. Anyhow, differently from any other technique, it is based on the administration of specifically designed game-based activities, which introduces a new kind of interaction with users in the repertoire of data gathering methods allowing use to overcome the limitation of traditional techniques: compared with questionnaires, it guarantees high quality of data; compared with interviews, it guarantees high quantity of data in short time, compared with user observation/field study, it prevents obtrusiveness since investigators do not interfere with routine activities but rather propose new ones. Furthermore, this method allows investigators to collect a high quantity of structured user-produced data (game results), to be archived for later (statistical) analysis. The data we gathered were qualitatively genuine (a child could express his/her true self) and dependable for creating fine-grained profiles of learners and their preferences. The reliability of data is supported by evidence from teachers and parents of the involved children (gathered via contextual inquiries). The new approach proved to be definitely engaging for children and teachers, to the point that the involved schools became so interested in the project that volunteered to participate in the prosecution of TERENCE activities (this allowed us to carry on a large scale evaluation with about 900 learners in two countries). The chosen approach also allowed us to conduct an extensive study with many users within time limit and organizational constraints. On the other hand it has be said that game design and game material constructions require considerable human resources, and that the semi-structuredness of collected data may make their analysis expensive. Notwithstanding these drawbacks, the attained results and their contribution to the success of the project make it reasonable to study if and how a game based approach can fit in the body of knowledge of UCD contextual studies, since its goals and effects may outbalance some flaws of traditional techniques not only for the new types of users that are entering the realm of technological artifacts, but also for more traditional users.

# References

1. Cofini, V., de la Prieta, F., Mascio, T.D., Gennari, R., Vittorini, P.: Design Smart Games with Context, Generate them with a Click, and Revise them with a GUI. Advances in Distributed Computing and Artificial Intelligence Journal 3 (December 2012)
2. Druin, A.: The role of children in the design of new technology. Behaviour and Information Technology (BIT) 21(1), 1–25 (2002)
3. Druin, A., Bederson, B., Boltman, A., Miura, A., Knotts-Callahan, D., Platt, M.: The Design of Children's Technology: How We Design, What We Design and Why. In: Children as Our Technology Design Partners, ch. 3. Kaufmann, M. (1998)
4. Guha, M., Druin, A., Chipman, G., Fails, J., Simms, S., Farber, A.: Working with young children as technology design partners. Communications of the ACM 48(1), 30–42 (2005)
5. Hanna, L., Risden, K., Alexander, K.: Guidelines for usability testing with children. Interactions 4(5), 9–14 (1997)
6. Jong, M.S., Lee, J., Shang, J.: Educational Use of Computer Games: Where We Are, and What's Next. In: Huang, R., Spector, J.M. (eds.) Reshaping Learning. New Frontiers of Educational Research, pp. 299–320. Springer, Heidelberg (2013)
7. Mascio, T.D.: First User Classification, User Identification, User Needs, and Usability goals, Deliverable D1.2a. Technical report, TERENCE project (2012)
8. Mascio, T.D.: Second User Classification, User Identification, User Needs, and Usability goals, Deliverable D1.2b. Technical report, TERENCE project (2012)
9. Mor, Y., Winters, N.: Design approaches in technology enhanced learning. Interactive Learning Environments 15(1), 61–75 (2007)
10. Przybylski, A.K., Rigby, C.S., Ryan, R.M.: A Motivational Model of Video Game Engagement. Review of General Psychology 14(2), 154–166 (2010)
11. Resnick, M., Martin, F., Berg, R., Borovoy, R., Colella, V., Kramer, K., Brian, S.: Digital manipulatives: new toys to think with. In: Proc of CHI 1998, pp. 281–287. ACM Press (1998)
12. Slegers, K., Gennari, R.: Deliverable 1.1: State of the Art of Methods for the User Analysis and Description of Context of Use. Technical Report D1.1, TERENCE project (2011)
13. Vaajakallio, K., Mattelmäki, T., Lee, J.: Co-design Lessons with Children. Interactions 17(4), 26–29 (2010)
14. Van Eck, R.: Digital game-based learning: It's not just the digital natives who are restless. EDUCAUSE Review 41, 16–30 (2006)

# Identifying Weak Sentences in Student Drafts:
# A Tutoring System

Samuel González López[1], Steven Bethard[2], and Aurelio López-López[1]

[1] National Institute of Astrophysics, Optics and Electronics, Tonantzintla, Puebla, México
{sgonzalez,allopez}@inaoep.mx
[2] University of Alabama at Birmingham, Birmingham, Alabama, USA
bethard@cis.uab.edu

**Abstract.** The first draft of an undergraduate student thesis generally presents deficiencies, which must be polished with the help of the academic advisor to get an acceptable document. However, this task is repeated every time a student prepares his thesis, becoming extra time spent by the advisor. Our work seeks to help the student improve the writing, based on intelligent tutoring and natural language processing techniques. For the current study, we focus primarily on the conclusions section of a thesis. In this paper we present three tutoring system components: Identifying Weak Sentences, Classifying the Weak Sentences, Customizing Feedback to Students. Our system identifies weaknesses in sentences, such as the use of general instead of specific terms, or the absence of reflections and personal opinions. We provide initial models and their evaluations for each component.

**Keywords:** Weak sentences, Thesis drafts, Conclusion evaluation.

## 1 Introduction

Writing a thesis involves complying with certain requirements and rules established by institutional guides of universities and authors of methodology books. In this way, students have guidelines to follow when developing their first draft. However, the experience of teachers is that the student theses present different types of errors, ranging from misspellings to content errors.

A survey applied to students of Computer-related careers about the problems in preparing a thesis revealed that 8% have problems with the structure of the document. This lack of knowledge leads students to write poor documents [1].

Our work focuses on the conclusions section of theses. In our current work, this section is considered acceptable if it provides: a) an analysis of compliance with each of the research objectives; b) a global answer to the research question, and c) a contrast or value judgment of the results. Its content should avoid the use of general terms instead of specific terms and the use of speculative words.

We develop a tutoring system that identifies sentences that do not follow these guidelines for the conclusion section. For instance: *In the project, we have developed*

T. Di Mascio et al. (eds.), *Methodologies and Intelligent Systems for Technology Enhanced Learning*, Advances in Intelligent Systems and Computing 292,
DOI: 10.1007/978-3-319-07698-0_10, © Springer International Publishing Switzerland 2014

*two concept tests, one has been to do the survey that collected data from people, and another has been to make a concept test of the peripheral[1]* Here, we can observe that the student describes a part of the experimentation, instead of providing a value judgment of the results. This sentence would be more suitable to another section of the thesis, for example, the methodology section. Our tutoring system starts by identifying weak sentences in the student conclusion. If the sentence is weak, the system identifies the type of weakness. Finally, the system sends feedback to the student depending on the type of weakness found. The results reported here are part of a larger project that can help students to evaluate their early drafts, and facilitate the review process of the academic advisor.

## 2     Background

An intelligent tutoring system (ITS) is a system that provides personalized instruction or feedback to students with the aim of giving them permanent support. Advances in ITS include the use of natural language technologies to perform automated writing evaluation and provide feedback.

This work [2] focuses on improving the cohesive devices of the student essay. Similarly, our work seeks to improve the writing, but focuses on sentences in the conclusion section. The tutor Writing Pal [3] offers strategy instruction and game-based practice for developing writers, and was also used with an experimental group, obtaining significant improvements in students who participated in the experiment. Our work considers such lexical, cohesion and rhetorical features, augmented with other measures with the goal of capturing behaviors specific to the conclusions section.

A dialogue-based ITS called Guru has an animated tutor agent engaging the student in a collaborative conversation that references a hypermedia workspace, displaying animating images significant to the conversation [4]. Another dialogue-based ITS, Auto Tutor, uses dialogues as the main learning activity [5]. One interactive essay-writing tutor was designed to improve science knowledge by analyzing student essays for misconceptions. The authors presented five modules that allow the system to identify if the student is using concepts improperly [6]. We take some inspiration from their work in developing our model of weak sentences. All these works use natural language to interact with the student, as we do.

## 3     System Overview

Our system has a Weak Sentences Identifier, which contains three main components (see Fig. 1). The component Identifying Weak Sentences (IWS) is responsible for discerning whether a sentence is weak or strong, and includes five models, that use different techniques and resources such as lexical richness, measure of similarity between sentences, the use of speculative terms and the overlap with terms from the conclusion sections of approved theses. The component Classifying the Weak

---

[1] All example sentences have been translated to English from the original in Spanish.

Sentences (CWS) looks at sentences identified as weak by the previous component and determines the kind of weakness. This component takes advantage of a corpus tagged by human annotators to train a model with the main weaknesses identified by them. The component Customized Feedback to Students (CFS) selects a paragraph of a conclusion section from an approved thesis to provide student support to improve his writing.

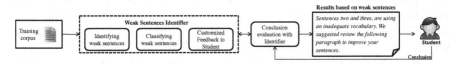

**Fig. 1.** System of Weak Sentences Identifier

Our corpus is focused on Spanish undergraduate and graduate theses in Computer Science. Currently we gathered 55 theses of Advanced College-level Technician degree (TSU)[2] level, took the 544 sentences in their conclusion sections, and sent those to human annotators to be tagged with strong and weak classes. The kind of research reported by students in their thesis is mostly using a quantitative approach.

Identifying a weak sentence in the conclusion is complicated for human annotators, since it requires expertise in computer science thesis advising and the ability to discern if a sentence in a conclusion complies with minimum requirements. Our annotators were instructors in Public Universities, teaching courses and advising students in their final semesters about their theses. The background of the annotators was in computing, specifically in information technologies. Annotators tagged each sentence as strong or weak. For sentences identified as weak, annotators provided the type of weakness. Annotators were provided with a guide defining the qualities of a good conclusion and including examples of weak and strong sentences. The criteria used by annotators to identify the two kinds of sentences, based on university institutional guides and authors of methodology books were: a global response to the research question, compliance with each of the research objectives, the acceptance or rejection of the hypothesis, and the contrast between the fundamentals and results.

We also collected 210 Bachelor and Master Degree theses. These documents were used by the component IWS after doing Unsupervised Clustering. The aim of the clustering was to identify sentences that are representative of concepts found in approved theses. This clustering was used only in the component IWS [7]. The component CFS also uses the corpus of Bachelor and Master degree to send suggestions for students, depending on the kind of weakness identified.

## 4    Identifying Weak Sentences

The models developed seek to capture features that reflect weak or strong sentences. A sentence presents weakness such as concepts written in general instead of specific terms, or the absence of reflections and value judgments. These types of sentences do

---

[2] A two-year study program offered in some countries.

not fit in the conclusion section. An example of a weak sentence, part of a thesis about computer networks, states: *"Security should not be a problem, neither for networks nor in everyday life, but as some humans do not have a social conscience either by greed, a bad curiosity, ambition"*. We notice in the sentence that the student is using speculative words such as "should". Also the student expresses a philosophic argument that fits better into an introduction section as a motivation of the problem.

A strong sentence means that the text is reasonable and makes sense in a conclusion. For instance, *"The new system will help to lower costs for man/hours invested in the maintenance of the infrastructure"*. We note in the sentence that the student is writing a possible consequence of the implementation of the system.

We used five models for identifying poor student sentences in the conclusion section. These are detailed next.

**Lexical Richness (LR)** is the first model and involves the *variety, density and sophistication* measures. The idea of this model is that if a sentence obtained low levels of lexical richness, then it was a weak sentence, otherwise, it was a strong sentence. Each measure was computed individually. *Variety* seeks to measure student ability to write with a diverse vocabulary, and is computed by dividing the number of content terms by the total of lexical types, both ignoring empty words (i.e. prepositions, conjunctions, articles, pronouns). *Density* aims to reflect the proportion of content words in the complete text, is intended to reveal an excessive use of empty words, and is calculated by dividing the number of content words by the total words of the evaluated text (i.e. before removing empty words). *Sophistication* attempts to reveal the handling of technical concepts, is the proportion of sophisticated words employed, and is computed as the percentage of words out of a list of 1000 common words from the Spanish Royal Academy. In [8], a pilot test with college students showed that students increased their scores and improved their texts, after the use of a lexical richness tool.

**Coverage Model (CM)**, the second model, uses the Core-Concepts (CC) obtained by MEAD[3], a tool that allows automatically extracting a set of CC from our corpus of graduate theses. CC are key ideas that support the learning of the student. In MEAD, each CC is represented by one sentence. In our work the CC represent strong sentences drawn from high quality, graduate thesis conclusion sections that can be used to identify weak sentences in TSU conclusions. An example of core concept extracted from our corpus is: *The Unified Modeling Language (UML) allows developers to make inroads into the paradigm of object-oriented design at analysis and design level, but not in implementation level.* We can identify that "UML" can be used with the paradigm of "object-oriented". A TSU conclusion that references similar topics can be given the paragraph containing this CC as an example. Such feedback can help them to improve their writing.

To use CCs in finding weak sentences, we employ a model similar to that of [6], counting the number of words in common between all CCs and the student sentence, obtaining a score. A low score means the sentence appears unlike any graduate thesis sentence and may therefore be a poor sentence. The score is calculated by dividing the

---

[3] http://www.summarization.com/mead/

intersection of the words of the student sentence and the CC domain (expanded with synonyms[4]) by the words of the student. The result is given in a range of 0 to 1, where a value close to 0 means that sentence is far from the CCs. We also explored a variant, **Coverage Model (CM2)**, which eliminates the empty (stop) words when comparing TSU sentences and CCs.

**Similarity of Cosine (SC)** With the goal of identifying sentences that do not fit with the rest of the sentences in the conclusion section, we computed the cosine of the sentence to be classified when compared to all sentences of the conclusion. The cosine was calculated treating each sentence as a vector of term counts. We define the SC score as the average of the all similarities calculated for the sentence to be classified. These steps were applied for each sentence. Therefore, we have averages for each sentence of the conclusion. A sentence with a low average is unlike all other sentences in its conclusion and may therefore be a weak sentence.

The last model developed regards the use of **Speculative Words (SW)** in the sentences by students. We take as reference the table of lexical features provided by [9] that includes modal auxiliaries (*may, might, could, would, should*), evidential verbs (*appear, seem*), adjectives (*likely, probable, possible*), adverbs (*probably, possibly, perhaps, generally*) and nouns (*possibility, suggestion*). We used the Spanish versions. The conclusions have to show evidence of reflections and the fulfillment of the objectives. If the sentences contain speculative words then the conclusion is anomalous. For example, the phrase "probably the results" shows that the student does not have certainty of results and this sentence may be a weak sentence. Also we develop a variant of this model, **Speculative Words Expanded (SWE)**, which includes colloquialisms and synonyms. The idea was to identify words that do not contribute value to the conclusion.

We evaluate our models to identify weak sentences with the corpus of TSU level. The first step was submitting the sentences of the conclusions to human annotators, with the aim of generating a gold standard. The sentences were tagged with the class "weak" and "strong", and for each "weak" sentence, a description of why it was considered "weak" was provided. A total of 165 sentences were tagged with weak class and 329 sentences with strong class. In Table 1, we show the confusion matrix of agreement and disagreement between annotators. Note that most agreements were on strong sentences.

**Table 1.** Confusion matrix between annotators

|  | *Weak* | *Strong* |
|---|---|---|
| *Weak* | 48 | 120 |
| *Strong* | 36 | 340 |

The Kappa-Cohen agreement between human annotators was of 0.25, corresponding to a Fair level of agreement. A third annotator adjudicated the disagreements between the two primary annotators. After annotation, we processed each sentence with the models to obtain features, and we used these results as input to

---

[4] Multilingual Central Repository, http://adimen.si.ehu.es/web/MCR/

a classifier. We used the Lexical Richness alone as baseline. Below we show the results obtained by the classifiers with 10-fold cross-validation. Since our goal is to remedy "weak" sentences, we are mostly interested in the precision and recall of the "weak" class, but we also show performance on the "strong" class.

Table 2 shows that all models outperformed the baseline F-measure of 0.527. We add to the baseline different model features with the goal of improving precision and recall. The system with the highest F-Measure for Weak sentences (0.622) was SC+SWE+CM2+LR, though SC+CM+LR achieved slightly higher precision (0.613 vs. 0.603). This may suggests that identifying speculative words generally improves recall, though at a small cost to precision.

**Table 2.** Classifying results

| Models | Precision | Recall | F-Measure | Class |
|--------|-----------|--------|-----------|-------|
| LR | 0.556 | 0.5 | 0.527 | Weak |
|    | 0.582 | 0.636 | 0.608 | Strong |
| SC+SW+CM+LR | 0.567 | 0.506 | 0.535 | Weak |
|             | 0.589 | 0.647 | 0.617 | Strong |
| SC+CM+ LM | 0.613 | 0.548 | 0.579 | Weak |
|           | 0.624 | 0.685 | 0.653 | Strong |
| SC+SW2+CM2+LM | 0.603 | 0.643 | **0.622** | Weak |
|               | 0.653 | 0.614 | **0.633** | Strong |

For our task, we seek a balance between precision and recall, like that of the SC+SWE+CM2+LR model, since the system has to clearly distinguish a weak sentence in the conclusion – otherwise the system could confuse the student – and at the same time should have good coverage.

## 5      Classifying the Weak Sentences

The goal of this component is to take the student sentences tagged as weak by the previous component and identify what kind of weakness the sentence shows.

We select all sentences that were tagged as weak (165 sentences) by human annotators to train a model. Annotators provided a description of why they thought a sentence was weak. We executed an unsupervised clustering over these descriptions with the objective of identifying which were the most common weaknesses of the corpus. We applied Latent Semantic Analysis clustering. Upon manually inspecting the clusters, we identified the following main types of weaknesses:

1. The sentence was not connected to research results (**NC**): The sentence does not show evidence of some kind of analysis or reflection of the results. Also, there is an absence of arguments to provide support for the results. For instance: *The strategy that was used for this project had good results.* We can see that the sentence does not show a contrast between strategy and results.

2. The sentence is written in General Terms (**GT**): The text is like an introduction, justification or related work. The sentence does not add value to the conclusion, only making it longer. For example: *This indicates that Ethernet will continue to evolve while other transmission technologies disappear.* Observe that the student fails to take their own position.

3. The vocabulary used in the sentence is not adequate (**VO**): This includes errors such as repetition of terms, repetition of words, and terms that are distant from the topic. Example: *Now, it is time to lay hold of all the tools that exist in our environment in order to economize on operations and do not forget to seek the security of this.* The student uses informal phrases like "lay hold".

Models useful for classifying the different kinds of weakness are: SW, SWE, CM, CM2, LR (described in previous sections). We evaluate our models for identifying the kind of weakness using as gold standard the "weak" class, tagged with the types we identified from the annotator descriptions: NC, GT and VO. There were 113 examples of GT, 34 of NC and 18 of VO. Because of the small number of sentences with types NC and VO, we merged them into a single NC_VO class. The results (Table 3) were obtained by classifiers using 10-fold cross-validation.

**Table 3.** Classification results

| Models | Precision | Recall | F- Measure | Class |
|---|---|---|---|---|
| SW+SWE+CM+CM2+Variety | 0.6 | 0.35 | 0.439 | NC_VO |
| | 0.748 | 0.894 | 0.815 | GT |

In this experiment the best combination of features was SW+SWE+CM+CM2+Variety. We tried other combinations that included density and sophistication (from the LR features) and cosine similarity models, but these did not perform as well. Sentence similarity features likely contribute less to this task because they are not focused on identifying vocabulary and term-based issues. In general, Table 3 shows that while the model's predictions of GT sentences are fairly reliable, identifying NC and VO sentences is more challenging, probably due to the small amount of training data available for these classes.

# 6     Customized Feedback to Students

The kind of feedback sent to students depends on the weakness type identified by the component CWS. For every weakness, the system will send a message to the student, showing as an example of a good conclusion, a paragraph from a Bachelor's or Master's degree thesis (which are of higher quality than TSU theses). The paragraph is intended to contrast with the identified problem in the TSU conclusion. For example, if the TSU conclusion sentence was classified as GT, then we want a paragraph that is not written in general terms.

This component takes as a reference the CCs that resulted from running MEAD on our corpus of Bachelor and Master degree. We seek to identify which of the CCs are related to each of the types of weaknesses identified. Thus we have two groups of

CCs: The first is related to the GT class and the second group is related to the kind of NC_VO class. Each group is ordered depending on how appropriate they are to correcting weak sentences with that class. Given a sentence that has been identified as weak and classified as NC, VO or GT, this component selects the most highly ranked CC of the appropriate class, retrieves from the corpus the paragraph containing the CC, and sends the paragraph as feedback. We expect that the paragraph helps the student to improve their conclusion. If the paragraph is not helpful, the system can continue down the ranked list to send additional feedback.

Models employed for feedback that are most useful for finding CCs to address the different kinds of weakness are: for NC_VO class we used SW, SWE, CM, CM2 and for GT class we used SC, SW, SWE, CM, CM2, and LR (all described above). The evaluation of our models consisted of identifying strong sentences that contrast with weak sentences of a particular type. We trained classifiers on the "strong" sentences plus just the "weak" sentences of a particular type. For example, we trained the classifier for responding to GT problems on the 329 "strong" sentences, plus the 113 GT sentences. Similarly, for responding to NC_VO problems, we trained on 329 "strong" sentences plus 52 NC_VO sentences. We then applied these two models to the sentences (CCs) from the Bachelor's and Master's degree theses, and ranked those sentences based on the score output by the classifier. With the goal of validating our schemes, we prepared a group of 212 ordered CCs and sent them to human annotators to evaluate the relevance of the CC with the weaknesses. This task is in progress and we are waiting for the CCs to be tagged, to validate our last component.

# 7    Discussion

In this paper, we have presented a system that uses natural language techniques and considers specific features of writing to help students improve the writing of thesis conclusion sections. This task was complex for human annotators, since it requires expertise in computer science thesis advising and discerning if a sentence in complies with the minimum requirements. In our work we take advantage of the knowledge of different academic advisors, who have annotated our corpus. We apply a variety of different models to characterize potential problem sentences in a conclusion, and use them to generate features for supervised classifiers.

We found that weak and strong sentences share features, i.e. a weak sentence contained similar terms as a strong sentence, but they had a different structure. These differences allowed the classifiers to identify patterns.

Moreover, we identified that the weak and strong sentences, labeled by the annotators, are not affected by the writing styles of each student. For instance the use of the active voice or passive voice in a conclusion does not affect the tagging process of sentences, since the criteria defined to label weak or strong sentences contemplated own aspects of a conclusion, such as contrast results. Also, it was identified that the weak and strong sentences, labeled by the annotators, are not affected by the writing styles of each student. For instance, the use of the active voice or passive voice in a conclusion does not affect the criteria that must satisfy a good conclusion.

However, there is still work to do. We need to increase the amount of training data for low frequency weakness types, e.g. inadequate vocabulary. This would allow our system to have better coverage of the different kinds of weaknesses, and to strengthen the features of weak and strong sentences. Furthermore, we could increase the number of annotators, to improve the level of Kappa-Cohen agreement, taking care of including annotators with a similar academic background, e.g. instructors with computer engineering degrees.

The set of components presented in this paper could be applied to other domains, such as the identification of weak sentences in essays of students learning English. Since many of the features are language independent, it would only be necessary to make some small number of changes in the text pre-processing and to use a corpus tagged by instructors with experience in reviewing English essays. As future work, we are planning to conduct an extrinsic test: a pilot study where students of TSU level write thesis conclusions and receive feedback, with the aim of verifying that our system helps students to improve their writing.

**Acknowledgments.** We thank the reviewers: J. Miguel García G., and Indelfonso Rodriguez E. This research was supported by CONACYT, México, through the scholarship 1124002 for the first author. The third author was partially supported by SNI, México.

# References

1. Muñoz, C.: Como elaborar y asesorar una Investigación de tesis. Prentice Hall (2011)
2. Crossley, S.A., Varner, L.K., Roscoe, R.D., McNamara, D.S.: Using Automated Indices of Cohesion to Evaluate an Intelligent Tutoring System and an Automated Writing Evaluation System. In: Lane, H.C., Yacef, K., Mostow, J., Pavlik, P. (eds.) AIED 2013. LNCS, vol. 7926, pp. 269–278. Springer, Heidelberg (2013)
3. Crossley, S., Roscoe, R., McNamara, D.: Using Automatic Scoring Models to Detect Changes in Student Writing in an Intelligent Tutoring System. In: Procs. 26th FLAIRS, pp. 208–213 (2013)
4. Olney, A.M., D'Mello, S., Person, N., Cade, W., Hays, P., Williams, C., Lehman, B., Graesser, A.: Guru: A Computer Tutor That Models Expert Human Tutors. In: Cerri, S.A., Clancey, W.J., Papadourakis, G., Panourgia, K. (eds.) ITS 2012. LNCS, vol. 7315, pp. 256–261. Springer, Heidelberg (2012)
5. Graesser, A., D'Mello, S., Craig, S., Witherspoon, A., Sullins, B., McDaniel, B., Gholson, B.: The Relationship between Affective States and Dialog Patterns during Interactions with Autotutor. Interactive Learning Research 19, 293–312 (2008)
6. Bethard, S., Okoye, I., Arafat, S., Hang, M.J., Sumner, T.: Identifying science concepts and student misconceptions in an interactive essay writing tutor. In: Procs. of the 7th Workshop on Building Educational Applications Using NLP, pp. 12–21 (2012)
7. Bellegarda, J.: Unsupervised document clustering using multiresolution latent semantic density analysis. In: Workshop on Machine Learning for Signal Processing, pp. 361–366 (2010)
8. García Gorrostieta, J.M., González López, S., López-López, A., Carrillo, M.: An intelligent tutoring system to evaluate and advise on lexical richness in students writings. In: Hernández-Leo, D., Ley, T., Klamma, R., Harrer, A. (eds.) EC-TEL 2013. LNCS, vol. 8095, pp. 548–551. Springer, Heidelberg (2013)
9. Kilicoglu, H., Bergler, S.: Recognizing speculative language in biomedical research articles: a linguistically motivated perspective. In: Procs. of the Workshop on Current Trends in Biomedical Natural Language Processing, Stroudsburg, PA, USA, pp. 46–53 (2010)

# Achieving Diagnostic Expertise through Technology: A Cue Based Approach

Barbara Giacominelli[1], Margherita Pasini[1], and Rob Hall[2]

[1] Università di Verona, Dipartimento di Filosofia, pedagogia e Psicologia,
Lungadige Porta Vittoria, 17; 37129 Verona
[2] Environmetrics Pty Ltd.,
PO Box 793, Pymble NSW 2073, Sydney, Australia
{barbara.giacominelli,marghertita.pasini}@univr.it,
rob@environmetrics.com.au

**Abstract.** Expertise is typically associated with high levels of experience in a domain. However, studies have shown that expertise is not necessarily correlated with experience [1]. The purpose of this paper is to describe a strategy for developing and evaluating a cue-based training program that can be delivered by computer. We describe how scenarios were developed to assess expertise amongst neonatal nurses and then evaluated for efficiency by treating the scenarios as test items and using the psychometric technique known as Item Response Theory.

**Keywords:** diagnostic expertise, decision making, cue-based theory, Item Response Theory, Learning.

## 1 Introduction

In the last 20 years many studies have focused on the cognitive processes used by experts when making decisions, in order to train decision-making skills and help people to achieve expertise. The framework adopted for the present study belongs to the Naturalistic Decision Making approach (NDM) [2-3]. The NDM approach investigates "the strategies people use in performing complex, ill-structured tasks, under time pressure and uncertainty and in a context of team and organizational constraints" [4]. Klein (1997) argues that expert and novice performance can be distinguished by the skills of acquiring, identifying and using cues for diagnosis and response. Cues can be considered features of the envi-ronment that, through knowledge and experience, have come to be associated, in memory, with particular events [5]. For example, the presence of dark clouds (feature) might be considered a cue, since there is an association with rain (the event). As a result of the presence of dark clouds, the decision-maker may be prompt-ed to take sufficient precautions to avoid exposure to rain.

Wiggins and O'Hare [6] demonstrated the efficacy of the cues-based approach in diagnosis by using a cue-based training system for weather-related decision-making. In their study participants using a computer-based flight simulator were exposed to a set of cues related to meteorological conditions that could have a bearing on their

T. Di Mascio et al. (eds.), *Methodologies and Intelligent Systems for Technology Enhanced Learning*, Advances in Intelligent Systems and Computing 292,
DOI: 10.1007/978-3-319-07698-0_11, © Springer International Publishing Switzerland 2014

flight planning. The authors concluded that if the cues that experts use to make a decision are known, it could be possible to construct a decision support sys-tem that would help less experienced operators to improve their decision-making capabilities. Subsequently, Wiggins and co-workers explored the influence of cues on the expertise of surgeons and nurses in a hospital context [5-1].

The cue-based studies reported by Wiggins et al use realistic scenarios as "test items" to identify which cues skilled and unskilled subjects refer to when making decisions about actions that need to be implemented under time pressure. Obviously, the validity of the method depends upon the ability of each scenario to discriminate between levels of expertise. In the study reported here, we developed a set of scenari-os in collaboration with some highly skilled neonatal nurses. We then used an approach based on Item Response Theory (IRT) to explore how well each scenario discriminated between a sample of neonatal nurses with various levels of experience and expertise.

## 2    Method

The study consisted in two steps: in the first step we developed three clinical scenarios and we used critical incident interviews [7] with 8 neonatal intensive care expert nurses to develop the clini-cal scenarios used in the present study. These nurses were selected by the supervisors of two different neonatal intensive care unit. The in-formation derived from the interviews was used to create 3 computer-based clinical scenarios. These consisted of a brief description of a pa-tient that described relevant symptoms and, where appropriate, details such as heart rate and temperature. The description was accompanied by instrument displays (such as heart monitors) that would be present in the ward. The second step of the study consisted in the testing of the clinical scenarios: for each of the three scenarios, participants were asked to review the information and then identify the abnormal reading as rapidly as possible. Each participant was tested alone but was moni-tored by the principal researcher.

### 2.1    Participants

The sample was 38 neonatal intensive care unit nurses (37 female, 1 male). They are all part of the nursing staff of a neonatal intensive care unit of a hospital. The participations was voluntary but all the staff took part to the study. They ranged in age from 24 to 59 years old with a mean of 36.8 years. The participants had worked in neonatal intensive care from between 0.5 to 40 years (mean period = 9 years; SD = 8.3 years).

### 2.2    Measures

The monitor display used in each scenario had one reading that had been judged abnormal by the experts. If a study participant identified this reading, they were judged to be "correct". The other choices were judged to be partially correct, in that they did provide some relevant clinical information, or incorrect.

## 2.3    Task Design

Critical incident interviews [7] were conducted with 8 neonatal intensive care expert nurses in order to develop the clinical scenarios used in the present study. In particular, the researcher asked the nurses to describe a situation in their job experience when they were successful and one when they were unsuccessful. Each interview was conducted using a semi-structured interview guide. The nurses were asked to remember the episode and describe it step by step. At each step they were also asked to identify which cues guided their decision and to rate how important each cue was to their decision-making. The information derived from the interviews were used to create 3 clinical scenarios, that corresponded at the 3 trials of the task presented at the nurses.

The particular type of task propose to the nurses, named Feature Identi-fication Task (FIT), is developed to train the initial recognition of a crit-ical condition and restrict the search for further cues [8]. Furthermore, this type of task requires the capacity to identify key cues from a complex scene, a skill that is presumably most devel-oped among experts [9-10] and is a known mediator of diagnostic performance [11]. In particular, participants were presented with a scenario with relevant cues and were asked to indicate as soon as possible which cue is most alarming and should inform a decision. This task simulate the nurses task in the real job environment. For the present study, a bedside monitor simulator was used to present cues[1].

FIT example, Scenario: *"Emma, 8 days old, arrives in neonatal* Task: *"As quickly as possible indicate the abnormal parameter on the bedside monitor"* (Figure 1 shows the monitor for this example).

Task: "As quickly as possible indicate the abnormal parameter on the bedside monitor" (Figure 1 shows the monitor for this example).

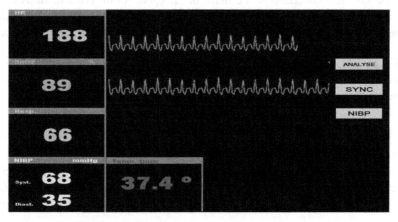

**Fig. 1.** Example of bedside monitor used for the FIT

---

[1] In this study we used an open source program for the vital signal monitor simulation named Healthy Simulation. Further information about this program is at this link: http://healthysimulation.com

The participant answers using a mouse and clicking on the abnormal parameter. For this study we presented the FIT in three trials, with three different scenarios and bedside monitors.

## 2.4    Procedure

Participants were initially briefed about the study in a meeting organized by the unit supervisor and were then asked to sign a consent form if they wished to participate. They subsequently provided some demographic details and then responded to the three scenarios presented on a laptop computer. The study was conducted in a dedicated room located in the neonatal intensive care unit and nurses did the experiment in their working hours.

# 3    Analysis

Two questions arise when a person takes a test of skill, knowledge or intelligence—how skillful (or knowledgeable or intelligent) is the person and how difficult is each of the items with which they were confronted in the test? Item Response Theory has been developed in the context of intelligence tests to tease apart these two elements that are confounded in a persons response to a any particular test item [12]. IRT is appropriate when it can be assumed that underlying a person's performance are two latent variables that are usually called "ability" and "difficulty". IRT models the data using a logistic function to describe the probability that a person will give a correct response as their ability increases. The shape of the logistic curve for an item tells us something about how well the item discriminates between people with higher and lower abilities. A comparison of the shape of the curve for different items as well as the parameter estimates of difficulty allow us to compare individual items (in contrast to an aggregated total "correct" score") and thus decide how useful each item in a test battery is. If we assume that the three scenarios in our study are test items that tap a single latent ability (skill or knowledge) dimension, then we can use IRT to assess how well out scenarios are functioning. We used the ITR module in the JMP (V11.1) statistical software to analyze our three scenarios.

# 4    Results

The three graphs of figure 2, at the end of the discussion, show the so-called Characteristic Curves for each of the three scenarios. The ability scale is arbitrarily centered around zero but allows for visual comparison between the three logistic curves.

The first scenario discriminated well between those people with an inferred higher level of ability. This is shown by the rapid rise in the graph indicating a switch from a low probability of being correct to a high probability. In contrast, the third scenario failed to discriminate well and the probability of choosing the correct element on the clinical display was relatively independent of ability. The second scenario was moderately good at discriminating. These findings can also be interpreted as meaning that scenario 1 was relatively difficult while scenario 3 was "too easy".

## 5     Discussion

We have presented here the first two elements of a strategy for developing and evaluating scenarios for use in a computer-based skill assessment tool. In practice, the development process can be iterative. For example, scenario 3 might be replaced by another scenario which would be, in turn, tested for efficiency. When a scenario such as scenario 1 has been shown to be relevant and able to discriminate efficiently between people of higher and lower skill levels, attention can turn to exploring the characteristics of the participants that are linked to making correct or incorrect responses. In particular, most prior studies of expert performance have identify job experts usually on the basis of experience in the domain [13-14]. Although these comparisons can be useful, they are based on the assumption that there is a linear relationship between experience and expert performance. In the next step of this research the task, linked to expert performance, could be used to investigate the relation between experience and expertise.

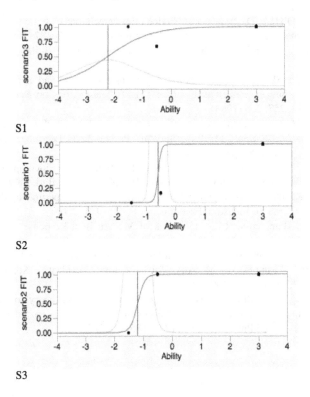

**Fig. 2.** Logistic curves of the ability scale in the three scenarios

**Acknowledgements.** The first author wishes to acknowledge the hospitality of Professor Mark Wiggins during her period in The Centre for Elite Performance, Expertise and Training at Macquarie University. She also wishes to thank Dr Thomas Loveday of the Centre for his advice and encouragement.

# References

1. Loveday, T., Wiggins, M., Searle, B.J., Festa, M., Schell, D.: The capability of static and dynamic features to distinguish competent from genuinely expert practitioners in pediatric diagnosis. Human Factors 55(1), 125–137 (2013)
2. Flin, R.: Distributed decision making for offshore oil platform emergencies. In: Proceedings of the Human Factor and Ergonomic Society 40th Annual Meeting, Pittsburg, PA. Human Factors and Ergonomics Society, Inc., Santa Monica (1996)
3. Zsambok, C., Klein, G.: Naturalistic Decision Making. Lawrence Erlbaum Associates Inc., Mahawah (1997)
4. Klein, G.A.: Developing Expertise in Decision Making. Thinking and Reasoning 3(4), 337–352 (1997)
5. Wiggins, M.: Cue-Based processing and human performance. In: Karwowski, I.W. (ed.) Encyclopedia of Ergonomics and Human Factors, pp. 3262–3267. Taylor and Francis, London (2006)
6. Wiggins, M., O'Hare, D.: Expert and Novice pilot perceptions of static in-fight images of weather. International Journal of Aviation Psychology 13(2), 173–187 (2003)
7. Flanagan, J.C.: The critical incident technique. Psychological Bulletin 51(4), 327–359 (1954)
8. Rasmussen, J.: Skills, rules, and knowledge: signals, signs, and symbols, and other distinctions in human performance models. IEEE Transactions on Systems, Man, and Cybernetics 13, 257–266 (1983)
9. Ratcliff, R., McKoon, G.: Sequential effects in lexical decision: Tests of compound cue retrieval theory. Journal of Experimental Psychology: Learning, Memory, and Cognition 21, 1380–1388 (1995)
10. Santos, P.J., Badre, A.N.: Automatic chunk detection in human-computer interaction. Paper presented at the Workshop on Advanced Visual Interfaces, Bari, Italy (June 1994)
11. Strauss, R., Kirlik, A.: Situation awareness as judgement II: Experimental demonstration. International Journal of Industrial Ergonomics 36, 475–484 (2006)
12. Linden, W.J., Hambleton, R.K.: Handbook of Modern Item Response Theory, vol. XV. Springer (1997)
13. O'Hare, D., Mullen, N., Wiggins, M., Molesworth, B.: Finding the Right Case: The Role of Predictive Features in Memory for Aviation Accidents. Applied Cognitive Psychology 22, 1163–1180 (2008)
14. Wallis, T.S.A., Horswill, M.S.: Using fuzzy signal detection theory to determine why experienced and trained drivers respond faster than novices in a hazard perception test. Accident Analysis and Prevention 39, 1177–1185 (2007)

# Learning Object Repositories with Federated Searcher over the Cloud

Fernando De la Prieta, Ana B. Gil, Antonio Juan Sanchez Martín, and Carolina Zato

Department of Computer Science and Automation Control, University of Salamanca,
Plaza de la Merced s/n, 37007, Salamanca, Spain
{fer,abg,anto,carol_zato}@usal.es

**Abstract.** The education sector is a significant generator, consumer and depository for educational content. Educators and Learners have access to technologies that allow them to obtain information ubiquitously on demand. The problems arising from the integration of educational content are usually caused by the vast amount of educational content distributed among several repositories. This work presents a proposal for an architecture based on a cloud computing paradigm that will permit the evolution of current learning resource repositories by means of cloud computing paradigm and the integration of federated search system.

**Keywords:** Learning Objects, digital repositories, information retrieval, cloud computing.

## 1 Introduction

The learning object paradigm is one of the main advances within the field of reutilization of educational resources. Formally, a Learning Object (LO) is defined [12] as *any entity, digital or non-digital, which can be used, re-used or referenced during technology supported learning.* In short, practically any educational resource (lesson, task, graph, subject, etc.), but there is a clear consensus that an LO must be the minimal reusable self-contained unit of learning content with a specific objective [2][14]. The paradigm is based in the fact of any education resource can be described by means for metadata, independently of its topic, format or size. The encapsulation of education resources in the form of metadata makes their digital distribution possible and therefore their reutilization, because this metadata allows making a first approach to the educational resource. The metadata schema is standardized. In fact, there are currently many standards such as DublinCore [5], IEEE LOM [12], etc. The existence of standards facilitates the management of the resources, enabling the interoperability among systems that use compatible standards.

LOs are stored in specific digital libraries, called Learning Object Repositories (LOR). Currently there is a significant growth of LOR as part of the hidden web in large databases. These systems typically provide a web interface to allow the

T. Di Mascio et al. (eds.), *Methodologies and Intelligent Systems for Technology Enhanced Learning*, Advances in Intelligent Systems and Computing 292,
DOI: 10.1007/978-3-319-07698-0_12, © Springer International Publishing Switzerland 2014

searching of education resources through the metadata. On of the main characteristic of these LORs is their heterogeneity [8] and therefore the interoperability among LORs is limited. However, to deal with this issue, they typically have a layer (interface) to makes possible the external access and hence, the interoperability. External search agent (a client or another LOR) can access. There are different standards or specifications that focuses on this interoperability layer, mainly OAI-MPH (*Open Archives Initiative for Metadata Harvesting*)[15] and SQI (*Simple Query Interface*) [6].

However, despite the theoretical advances done within this paradigm, the reality shows that its implantation in the real live is still limited [11]. From our point of view, there are two main problems. Firstly, from the usability viewpoint, the data that the authors assign to each descriptor of the metadata (independently of the specific standard) is very important, because this data that is used for searchers and if it is not correct, the results of the searches will be incoherent. To this end, it is necessary to follow a traceable process from the creation of an educational resource to the creation of its metadata in order to establish a metadata structure that is consistent, relevant and interpretable. However, the existence of many standards, the interoperability among them, the difficult to use the authoring tools and style of explanations by the authors, exacerbate the problem. And secondly, from a technological point of view, the heterogeneity of the repositories and their malfunction (as it is shown in the following section) constitute one of the main weakness of the paradigm.

This work presents the evolution of platform CLOR [3], by means of the integration of the federated searcher AIREH [9] in order to produce a clear advantage in the context of LO paradigm. On the one hand, CLOR (*Cloud-based Learning Object Repository*) is the present of a new generation of LOR because it is deployed into a cloud platform and makes use of the advantages of this computational paradigm (non-SQL databases, unlimited storage, etc.). On the other hand, AIREH (*Architecture for Intelligent Retrieval of Educational content in Heterogeneous Environments*) is a platform that makes possible the federated searches among many LORs and it integrates a recommender system [13].

This paper is structure as follow next section includes an study of the state of the art of current LOR and a real evaluation. Section 4 presents the proposed system and, finally, last section includes the results and conclusions.

## 2    Open Issues – Learning Object Repositories

LOs are commonly stored in repositories, which are characterized by their heterogeneity. The deployment infrastructure can basically be either distributed or centralized. Taking into account that an LO is formed by a digital resource and its metadata, there are four kinds of possible infrastructures [8]: (i) centralized resources and centralized metadata, (ii) centralized resources and distributed metadata, (iii) distributed resources and centralized metadata and (iv) distributed resources and distributed metadata. Furthermore, three kinds of storage strategies can be distinguished [8]: (i) *File-based,* which uses files with predefined formats and an index-based management; (ii) *Database-based,* which uses any kind of database, and is the most extended method; and (iii) *Persistent objects-based,* where the LO are stored as serialized objects.

The main problem is that LORs still do not implement any abstraction layer that can encapsulate the internal logic of the repository. Consequently, the search process and LO harvesting is a slow process, which require the manual intervention of users who must reuse the learning resources.   In this sense, to isolate this internal heterogeneity (storage techniques and models), there are interoperability layers which serve as a middleware layer between the repository and the clients, are (i) OAI-MPH (*Open Archives Initiative Protocol for Metadata harvesting*) [15] which is a protocol that provides a technology-independent framework for retrieving documents or resources, thus enabling interoperability among systems; and (ii) SQI (*Simple Query Interface*) [6] that is formed by a set of abstract methods based on web services. SQI is also is neutral in terms of the format of results as well as query language. This interfaces supports synchronous/asynchronous and stateful/stateless queries.

The state of the art shows a high heterogeneity in existing standards. Therefore, a study of LOR has been performed in order to analyze the real situation. In general, the systems in which this layer is included, suffer from various problems such as:

- The problems associated with the monolithic structure of LOR, which does not allow external management with the flexibility and power necessary to ensure easy interoperability, and dispersed and heterogeneous sources.
- the absence of automatic mechanisms that control the technical quality, semantics and syntax of LO, ensuring the correct specification of such LOs in any of the metadata schemas that describe them.

The study includes the analysis of the following LORs: *Acknowledge, Agrega, Ariadne, AriadneNext, CGIAR, EducaNext, LACLO-FLOR, LORNET, MACE, Merlot, Nime, OER Commons* and *Edna Online*. It consists of performing 60 queries to each LOR through an SQI layer that the repositories provide.  All of them use IEEE LOM [12] as metadata schema and VSQL [1717] as query language. Additionally, the majority of them are stateless (65%), and all of them have synchronous interfaces, but only 4 have the asynchronous interface.

The test shows that 6 of the 14 repositories do not work or are unavailable and they have to be removed from the scope of this study (Ariadne, AriadneNext, EducaNext, Nime and EdNa Online). MACE and LOCLO-FLOR produce an error in the authentication. After this step, this test is reduced to only four repositories Acknowledge, Agrega, LORNET and Merlot. The latter three are perfectly valid and all SQI methods work perfectly; however the repository Acknowledge only implements the essential methods to perform queries (Note that SQI specification does not force the implementation of all methods of the specification).

The graph on Fig. 1 shows the average of the time response a federated search is performed to these four alive LOR, the total average in the 60 queries is 3,718 seconds. The following graph in Fig. 2 shows the number of results retrieved per federated search during the test. The average of results is 24,24 result/querie.

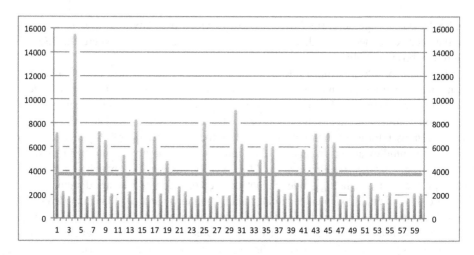

**Fig. 1.** Time average of a federated search in 4 repositories

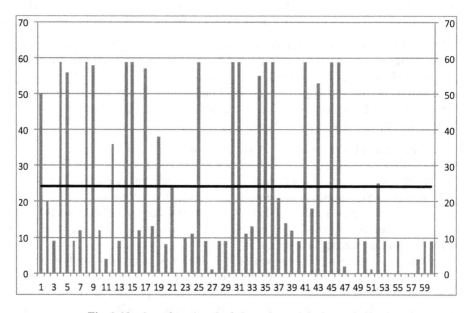

**Fig. 2.** Number of results of a federated search in 4 repositories

As it is possible to observe, the performance of the LOR is not appropriate. In order to deal with this problem, new LOR architectures have to be proposed and developed. This new generation of LOR must ensure the availability of resources and interoperability, permitting federated searches from external clients.

## 3    Proposal Architecture

These problems require solutions that are adapted to the heterogeneity. The solution should enable a centralized global search and the effective reuse of resources by the end user. As it is advanced on the introduction of this work is the integration of AIREH and CLOR in order to establish a system that not only allow the federated search among several LOR, but also the retrieve and safe storage of the retrieved LO.

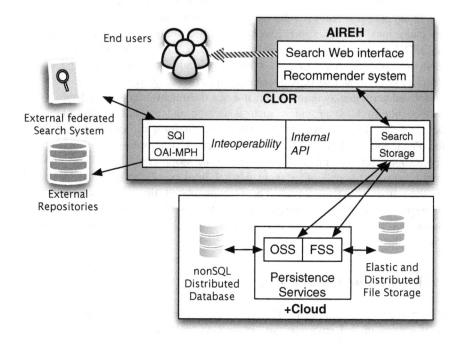

**Fig. 3.** CLOR and AIREH integration

The components of system, which are showed in Fig. 3, are described as follow:

— AIREH unifies the search and retrieval of objects, thus facilitating the learning search process by filtering and properly classifying learning objects retrieved according to some rules. The framework is based of a virtual organization [10] of intelligent agents that allow to deal with the heterogeneity of the environment. This new architecture will solve the problems of distribution, the integration of different repositories, the abstraction of the internal logic of each repository, and the classification, storage and retrieval of LOs, in a completely transparent way. In addition to adding capacities, such as simple scalability, to possible situations involving the use of new protocols, the architecture also adds internal logical repositories, cataloging or heterogeneous applications designed to cover services related features. AIREH implements Case Based Reasoning (CBR)[7] [1] that uses previous search information to rank the items that best suit the needs of the application user based on previously obtained information. It uses the profile information of each user as well as their educational information (content-based filtering).

— CLOR provides to AIREH the capacity to store the profile of each user as well as the persistence of retrieved LO (not only the metadata, but also the education resource). It is complemented with different interoperability layers, such as SQI or OAI-MPH, which will ensure the communication with other LORs and federated searches from external clients. It is framed at the platform level within Cloud services. Its main task is to encapsulate the communication with the lower layers of the Cloud platform that provides the need computational resources for the storage the educational resources in the web service file storage system and the metadata (in JSON format) associated with each resource into a non-SQL database. The main advantage is that it permits storing any kind of metadata independent of its structure or schema, that is, its standard. Furthermore, queries about the LO will be performed very quickly thanks to the use of a document-oriented database [16].

— +Cloud platform [4] provides, such as storage and databases. This platform is based on the Cloud Computing paradigm. This platform allows offering services at the PaaS and SaaS levels. The SaaS layer is composed of the management applications for the environment (virtual desktop, control of users, installed applications, etc.), and other more general third party applications that use the services from the PaaS layer. The components of this layer are the identity Manager, a File Storage System base on Web services, and an Object Storage Service ch provides a simple and flexible schemaless data base service oriented towards documents.

## 4    Results and Conclusions

The proposed system has been evaluated by performing a battery of tests to validate their efficiency in real environments. Evaluation metrics from information retrieval fields were adopted. The two most commonly used evaluation measures are precision (the fraction of documents retrieved by the system that are also relevant to the query) and recall (the fraction of the relevant documents present in the database that are retrieved by the system). Fig. 4 shows the number of relevant LO retrieves per query in comparison with other repositories, and Fig. 5 shows an average of time expended in retrieve LO per query.

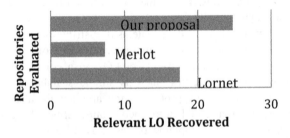

**Fig. 4.** Relevant LO recovered per query

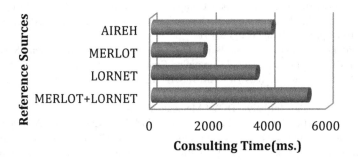

**Fig. 5.** Average time of queries

The first problem of LO paradigm are the incoherence in the medatada because the labeling process, which is basically done by hand, generates documents with serious shortcomings as there are many deficiencies related to the lack key attributes in the description. This makes it difficult, or impossible in some cases to study this aspect.

The second problem is the heterogeneity of the repositories and their malfunction. The proposed system tries to minimize this second problem, because it deals with the existing open issues:

1. The proposed model allows dealing with the heterogeneity of current and future standards since it is based on a non-relational database.
2. Cloud computing paradigm allows offering services with the same level of quality independently of its demand.
3. The low linkage among components permits implementing many interoperability layers without needing to upgrade to other modules.
4. The integration of AIREH allows filtering the results according to certain parameters related to the quality of the retrieved metadata documents; the user context information is then integrated into the use of the LOs.

**Acknowledgements.** This work is supported by the Spanish government (MICINN) and European FEDER funds, project iHAS: Intelligent Social Computing for Human-Agent Societies (TIN2012-36586-C03-03).

# References

1. Bajo, J., Corchado, J.M.: Evaluation and monitoring of the air-sea interaction using a CBR-Agents approach. In: Muñoz-Ávila, H., Ricci, F. (eds.) ICCBR 2005. LNCS (LNAI), vol. 3620, pp. 50–62. Springer, Heidelberg (2005)
2. Chiappe, A., Segovia, Y., Rincon, H.Y.: Toward an instructional design model based on learning objects. Educational Technology Research and Development 55, 671–681 (2007)

3. de la Prieta, F., Bajo, J., Marín, P.A.R., Méndez, N.D.D.: A New Generation of Learning Object Repositories Based on Cloud Computing. In: Casillas, J., Martínez-López, F.J., Vicari, R., De la Prieta, F. (eds.) Management Intelligent Systems. AISC, vol. 220, pp. 99–106. Springer, Heidelberg (2013)

4. De la Prieta, F., Rodríguez, S., Bajo, J., Corchado, J.M.: A Multiagent System for Resource Distribution into a Cloud Computing Environment. In: Demazeau, Y., Ishida, T., Corchado, J.M., Bajo, J. (eds.) PAAMS 2013. LNCS, vol. 7879, pp. 37–48. Springer, Heidelberg (2013)

5. Dublin Core Metadata Initiative. DCMI Metadata Terms

6. European Committee for standardization – Cen Workshop Agregament. A simple Query Interface Specification for Learning Repositories. Ref. No.: CWA 15454:2005 E (2005)

7. Fdez-Riverola, F., Corchado, J.M.: CBR based system for forecasting red tides. Knowledge-Based Systems 16(5), 321–328 (2003)

8. Frango, I., Omar, N., Notargiacomo, P.: Architecture of Learning Objects Repositories. Learning Objects: standards, metadata, repositories & LMS, pp. 131-155 (2007)

9. Gil, A.B., De la Prieta, F., Rodríguez, S.: Automatic Learning Object Extraction and Classification in Heterogeneous Environments. In: Pérez, J.B., Corchado, J.M., Moreno, M.N., Julián, V., Mathieu, P., Canada-Bago, J., Ortega, A., Caballero, A.F. (eds.) Highlights in PAAMS. AISC, vol. 89, pp. 109–116. Springer, Heidelberg (2011)

10. Gómez-Sanz, J.J., Pavón, J., Garijo, F.: Meta-models for building multi-agent systems. In: Proceedings of the 2002 ACM Symposium on Applied Computing, pp. 37–41 (2002)

11. Hodgins, H.W.: The future of learning objects. Educational Technology-Saddle brook then Englewood cliffs NJ 46(1), 49 (2006)

12. IEEE Learning Objet Metadata (LOM). Institute of Electrical and Electronics Engineers (2002), http://ltsc.ieee.org

13. López, V.F., de la Prieta, F., Ogihara, M., Wong, D.D.: A model for multi-label classification and ranking of learning objects. Expert Systems with Applications 39(10), 8878–8884 (2012)

14. Lujara, S.K., Kissaka, M.M., Bhalaluseca, E.P., Trojer, L.: Learning Objects: A new paradigm for e-learning resource development for secondary schools in Tanzania. World Academy or Science, Engineering and Technology, 102–106 (2007)

15. Lagoze, C., Van de Sompel, H., Nelson, M., Warner, S.: The Open Archives Initiative Protocol for Metadata Harvesting. Open Archives Initiative (2002)

16. Oren, E., Delbru, R., Catasta, M., Cyganiak, R., Stenzhorn, H., Tummarello, G.: Sindice. com: a document-oriented lookup index for open linked data. International Journal of Metadata, Semantics and Ontologies 3(1), 37–52 (2008)

17. Simon, B., Massart, D., Van Assche, F., Ternier, S., Duval, E.: Authentication and Session Management. Version 1.0. (2005)

# Psychometric Tests in the Field of Drawing Based in Timing Measurements

C. Lopez[1,*], F. Lopez[1], L.F. Castillo[3,4], M. Bedia[2], T. Gomez[2], and M. Aguilera[2]

[1] Department of Plastic and Physical Expression, University of Zaragoza, Spain
[2] Department of Computer Science, University of Zaragoza, Spain
[3] Department of Engineering Systems, National University of Colombia, Colombia
[4] Department of Systems and Informatics, University of Caldas, Colombia
c.lopez@unizar.es

**Abstract.** In the Art Education studies, one of the most interesting trends - in terms of its theoretical foundations- is the phenomena of measuring aesthetic experiences [1],[2]. However, traditionally, the teaching of art has been focused into the technical dimension of the drawings (proportionality, composition, etc.) or in cultural aspects (symbolic role, social criticism, etc.) but the analysis of aesthetic experiences have not acquired such an important status to be scientific analyzed and included in a pedagogical domain. In this paper we propose that, through metrics able to detect the 'immersion" or "sensitivity" of an artist drawing, it would be possible to get methodological tools capable of measuring aesthetic experiences. In order to explore our hypothesis, we have analyzed the work of students of art in several experiments and we have recorded the traces of the their drawings. The traces have been evaluated and we have tested if fractal timing is found (fractal timing is a ubiquitous clue that indicates that the among the components of a cognitive system exist a coupling). The results of the experiments seem to provide a way to improve some aspects in the daily practice of art teaching and a new way to describe psicometrical parameters.

## 1   Introduction

Student learning in the arts can be viewed in different ways. However, for the purposes of this paper, learning will be only used as "acquiring skills in one or more art forms". From this perspective, we are not going to refer to student content knowledge such as, for example, the history of art or the cultural relevance of art. In contrast, art skills refers to the abilities that allow students to produce art (for example, such as how well students perform a portrait). In this specific domain, teachers traditionally say that one of the main problems for evaluating artistic skills is the lack of assessment tools able to capture artistic attributes that (we think) remains in the domain of the subjectivity. How, for example, can we measure the aesthetic experience?

---

[*] Corresponding author.

T. Di Mascio et al. (eds.), *Methodologies and Intelligent Systems for Technology Enhanced Learning*, Advances in Intelligent Systems and Computing 292,
DOI: 10.1007/978-3-319-07698-0_13, © Springer International Publishing Switzerland 2014

There exist a discipline, psychometrics, devoted to create models and instruments to quantify aspects of our cognition. It is considered the field of the "psychological measurement" and fundamentally is concerned with the definition and validation of tools such as questionnaires, tests, or cognitive assessments [3]. It is well-known that the most famous psychometric tests were designed to measure the concept of intelligence (the Stanford-Binet IQ test), but, however, in the field of art education, psychometric instruments are almost nonexistent. It is difficult to measure attributes as "creativity". While terms as "general intelligence" can be modeled as the ability to solve problems, creativity is more related with the ability to open problems where one can not easily define which answer is correct. These complications have led to psychometric measurement of creativity and, in general, art studies show some "delays" compared to other aspects of intelligence [4]. There have been interesting proposals (for example, the Torrance Test [5]), which consists in performing some tasks such as telling stories from a drawing or enumerate the consequences that would occur after the occurrence of an unusual event) but they are exceptions.

In this paper we propose a new approach to measure and validate aspects related to the phenomena of artistic immersion. In order to get it, we need to show the notions on which our proposal is supported. In section 2, we review the scientific and philosophical literature regarding to "hand-tool" couplings [7] and, in section 3, we check some relevant experiments of how to measure the types of couplings mentioned [11] before that we propose, in section 4, how to apply those ideas in the domain of drawing. The promising results (in section 5) open the possibility to get a method to calibrate and quantify the artists experience.

## 2    Measuring Timing in the Artistic Skills

Although it is said that "beauty only may be in the eye of the beholder", it is known that the eye of the beholder has a biological preference based on mathematical proportions that can be measured. For example, within the context of the artistic man-made environment (sculpture, painting, architecture, etc.) there is a exhaustive preference for "golden section proportions". In contrast, in this paper we are interested in studying the possibility of defining a type of analogous characteristic of beauty but in a temporal dimension. We are interested in understanding how it could be one way to model and measure how people feel "artistic beauty" along a process. That is, how would it be a time-based equivalent of the golden rectangle? What would it be the "mathematical proportions" that determine "artistic values" across time?

### 2.1    The Classical Notion of "Ready-to-Hand" Revisited

The philosopher Heidegger argued that the ontological structure of the world is not given but arises through interaction [7]. With the philosophical expression "present-at-hand" (in contrast to ready-to-hand) it is referred an attitude like that of a theorist who merely is observing something. In seeing an entity as present-at-hand, the beholder is concerned only with the concept and prepared to theorize about it. But, however, presence-at-hand is not the way things in the world are usually encountered (really, this mode is only revealed as a secondary mode). In almost all cases we are involved in our world in an ordinary, and more involved, way, that is defined as "ready-to-hand" mode. Take for example, a hammer: it is ready-to-hand, if we use it without theorizing. Only when it breaks or something goes wrong might we see the hammer as present-at-hand.

These philosophical ideas are much more related to the computer science than one would imagine at first. For example, they are not entirely new into human-computer interfaces domains and interests. It was one of the elements on which Winograd and Flores [6] based their analysis of computational theories of cognition. It is easy to see the usefulness in examples like the following. Consider the way we use a mouse connected to a computer. Much of the time, we "act through the mouse"(i.e., the mouse is an extension of our hands when we select objects, operate menus, etc.). In this situations, the mouse would be defined as ready-to-hand. Sometimes, however, if we cannot move the mouse further, we become conscious of the mouse mediating my action, and the mouse, in fact, becomes in the object of our attention. It is through this moment of becoming "present at hand" that the object takes on an existence as an entity. For our interests, what is relevant about these notions, is that these terms have not been left alone in a philosophical domain but that some experiments have been used to quantify and measure this change of mode. In the next section, we will see the most relevant results.

## 2.2   Characterizing Hand-Tool Couplings

Several results along this decade has shown that fractal noise [8],[9], also known as fractal timing, is a signature of cognitive abilities when the cognitive task is generated by a softly assembled system. Research on the role of fractal noise in cognition has allowed a series of interesting results in the last decades in cognitive science. For example, in [10] the authors modeled the "insight" situation (the "aha" moment) in problem solving and they found that it coincides with a temporary increase appearance in fractal noise measured in hand and eye movements [11]. Following these works on fractal signatures, some experiments were designed in order to see whether the concept of "ready-to -hand" is related to fractal noise clues [8]. The experiments were organized using a virtual framework able to track the hand movements of people using a mouse to guide a cursor during a series of motor tests. When authors of [11], [12] analyzed how people moved the mouse, they found profound differences between patterns produced during mouse function and malfunction, and they also found relationships with fractal noise measurements. When the mouse worked, hand motions followed a mathematical form of fractal noise. But when the researchers mouse malfunctioned, the fractal noise vanished. Computer malfunction made test subjects aware of mouses and the computer was no longer part of their cognition. Only when the mouse started working again then concentration returned to normal level. These results demonstrated how people fuse with their tools, and, more important, they make significant the use of fractal noise as a measure of the ability of living the immersive experience of being coupled to a tool.

As we have mentioned, in both experiments subjects were required to perform a simple task using a tool and, although its true that the experience of immersion always remains subjective, it is also true that these empirical results provided a different point of view about the hand-tool coupling phenomena. Our question here is, Can these measures, that detect effortless and automatic tasks, be a way to measure aesthetical experiences?

## 3    Experimental Test to Measure the Artistic Skills

### 3.1    Participants and Procedure

In order to evaluate our hypothesis, some experiments with students of the School of Arts at Zaragoza (Spain) were organized and approved. All participants (45 undergraduate students, 22 female, 23 male, mean age = 19.4, range = 18-22) signed informed consent forms.

At the beginning, the experimenter explained that the nature of the task was to investigate artistic skills in the context of drawing with a electronic pen (able to record the traces of a drawing, to capture pressure-sensitive, and capture every nuance of the movements with high level precision). Students were recorded by drawing with the right hand (dominant hand) and the left hand. It is assumed that they will be able to be more concentrated when drawing with the dominant hand. Regarding to the task, the specific instructions given were to try to draw as concentrate as possible.

Procedure and Tasks: Several stimuli were presented in two experimental tasks: right-hand drawing, left-hand drawing. Once the participant have understood the rules of the experiment, four experimental trials (four different drawings) were performed. The same experiment but with the no-dominant hand, took place below. Participants were subjected to the same conditions as above.

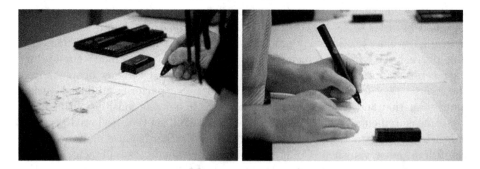

**Fig. 1.** Picture of the participants in the experiment: (a) right hand and (b) left hand

```
<?xml version="1.0" ?>
<ink>
    <definitions>
        <traceFormat id="default">
            <channel name="X" min="-3000" max="3000" units="0.1mm" />
            <channel name="Y" min="0" max="6000" units="0.1mm" />
            <channel name="Pressure" min="0" max="10000" units="none" />
            <channel name="xtilt" min="-127" max="128" units="none" />
            <channel name="ytilt" min="-127" max="128" units="none" />
            <channel name="timestamp" min="0" max="128" units="0.1sec" />
        </traceFormat>
        <context id="default-context" traceFormatRef="#default" />
    </definitions>
    <tracegroup id="layer0">
        <trace contextRef="#default-context">269.4 459.9 0 268 0 0 0,269.4 459.9 0 268
        <trace contextRef="#default-context">256.1 446.5 0 51 59 36 0,255.6 444.5 0 51
        <trace contextRef="#default-context">284.5 445.5 0 166 52 37 0,283.2 443.9 0 16
        <trace contextRef="#default-context">289.8 457.3 0 153 52 36 0,289.1 456.1 0 15
        <trace contextRef="#default-context">297.2 447.3 0 232 46 44 0,296.7 446.1 0 23
```

**Fig. 2.** Data generated by the experimental tool. Results of pressure, tilt, and cartesian coordinates are extracted.

## 3.2  Apparatus

For the experiments, we use Wacom's pens [13], with 2048 levels of pen pressure sensitivity, and the possibility of recording traces and tilt recognition. They allow us to reproduce with precision the movements of the student's hands while drawing. Figure 3 shows the type of data that can be obtained to analyze the movement of the hand.

## 3.3  Analysis

When we find fractal noise in the analysis of a signal, it is referring to a phenomenon of the spectral density, S(f) in a stochastic process, where f is the frequency on an interval [9]. Fractal noise is an intermediate situation between the well understood white noise (with no correlation in time) and random walk (or brownian motion noise with no correlation between increments). The experiments were analyzed using Detrended Fluctuation Analysis (DFA), a technique which allows us to estimate a coefficient of temporal correlation in a time serie [14].

For our analysis, we have to transform the data into appropriate variables. Only displacement of the pen in one coordinate (x or y) was used because a one-dimensional time series is sufficient for applying DFA and similar techniques, and to obtain the acceleration (the parameter of interest in our analysis). We consider, as has been explained in [11], that acceleration series is the relevant variable here because it corresponds to the active control ("the intentional purpose") on the part of the participant. Once obtained these data, they are organized as a function of the length $N$ of the segments. So, we will obtain a series of fluctuation functions F(n) as a sequence of consecutive blocks with size $n$. In

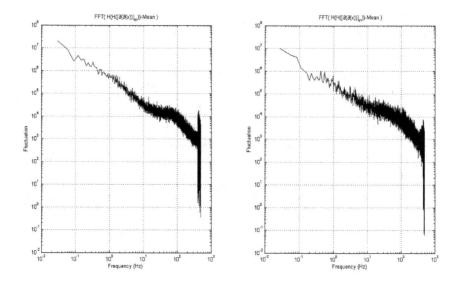

**Fig. 3.** Fluctuation functions from the behavior with dominant-hand (a) and (b) non-dominant-hand. It is shown that effortless and automatic tasks are related to immersive artistic skills.

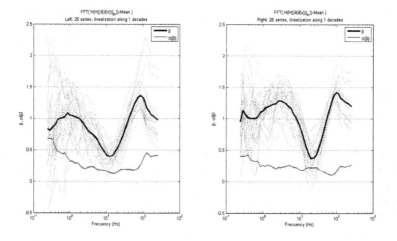

**Fig. 4.** Inflection point around 20-30 Hz for the right hand and around 10-20 Hz for the left hand

a log-log plot, the Figure 3 shows the exponent in the power-law relation $F(n)$ that corresponds to the scaling parameter of the fractal noise. In the special case when the exponent is close to 1, then it means that the task was made in immersive conditions [11]. Otherwise, when we find values of a significantly lower than 1 indicates the presence of disruptions. In our case, these situations seem to be connected also with low level inmmersive situations.

In Figure 4 (Graph fluctuations) a detailed observation reveals an inflection point around 20-30 Hz for the right hand and around 10-20 Hz for the left hand. This turning point can be seen more visibly to represent the exponent-coefficient of fractal noise (averaged over a decade, eg. 0.5 to 5 Hz to 1.58 Hz, coefficient since 1.58 is the midpoint in logarithmic scale), which shows clearly a minimum value around 20-30 Hz for the right hand and around 10-20 Hz for the left hand.

### 3.4   Conclusions

Our purpose has been to show an overview of a way to explore, develop and go beyond the weaknesses of the current psychometric approaches in order to study and evaluate artistic skills. We have shown that major areas of psychometric work have examined intelligence as problem-solver ability, but however, few work has been developed in Art Education. The experiments described above have two important implications in the artistic domain and in pedagogical aspects. First, experiments give support to our hypothesis about the use of fractal analysis to characterize artistic skills in drawing. Second, students also were consulted about its subjectivity and confirmed that those tasks that are more immersed, they are associated with feelings more aesthetical. Finally, the implications of these ideas should be tested in future studies, both psychometrically and through other methodologies.

## References

1. Dewey, J.: Knowing and the known. Beacon Press (1949)
2. Shusterman, R.: Pragmatist Aesthetics: Living Beauty, Rethinking Art, 2nd edn. Rowman & Littlefield Publishers (2000)
3. Furr, M., Bacharach, V.: Psychometrics: An Introduction. SAGE Publications, Inc. (2007)
4. Arden, R., Chavez, R.S., Grazioplene, R., Jung, R.: Neuroimaging creativity: A psychometric view. Behavioural Brain Research 214(2) (2010)
5. Hee Kim, K.: A Review of the Torrance Tests of Creative Thinking. Creativity Research Journal 18(1), 3–14 (2006)
6. Winograd, T., Flores, F.: Understanding computers and cognition: A new foundation for design. Ablex Publishing Corporation, New York (1986)
7. Dreyfus, H.: Being-in-the-world: A commentary on Heidegger's Being and Time, Division I. The MIT Press, Cambridge (1991)
8. Van Orden, G., Holden, J., Turvey, M.: Self-Organization of Cognitive Performance. Journal of Experimental Psychology: General 132, 331–350 (2003)
9. Holden, J., van Orden, G., Turvey, M.: Dispersion of response times reveals cognitive dynamics. Psychological Review 116, 318–342 (2009); van Orden, G., Holden, J., Turvey, M.: Human cognition and 1/f scaling. Journal of Experimental Psychology: General 134, 117–123 (2005)
10. Stephen, D.G., Boncoddo, R.A., Magnuson, J.S., Dixon, J.A.: The dynamics of insight: Mathematical discovery as a phase transition. Memory & Cognition 37, 1132–1149 (2009)

11. Dotov, D.G., Nie, L., Chemero, A.: A demonstration of the transition from ready-to-hand to unready-to-hand. PLoS ONE 5(3), e9433 (2010), doi:10.1371/journal.pone.0009433
12. Nie, L., Dotov, D., Chemero, A.: Readiness-to-hand, extended cognition, and multifractality. In: Carlson, L., Hoelscher, C., Shipley, T.F. (eds.) Proceedings of the 33rd Annual Meeting of the Cognitive Science Society, pp. 1835–1840. Cognitive Science Society, Austin (2011)
13. Interactive Pen Displays & Digital Drawing Tablets, http://www.wacom.com
14. Peng, C., Mietus, J., Hausdorff, J.M., Havlin, S., Stanley, H.E., et al.: Long-range anticorrelations and non-Gaussian behavior of the heartbeat. Physical Review Letters 70, 1343 (1993)

# Evaluation of Interactive Children Book Design
## The Case Study of "Little Rooster"

Zsófia Ruttkay, Judit Bényei, and Zsolt Sárközi

Moholy-Nagy University of Art and Design, 1121 Budapest, Zugligeti út 9, Hungary
{ruttkay,benyei}@mome.hu, zsolt.sarkozi@gmail.com

**Abstract.** In spite of mushrooming of interactive books apps for kids, there is hardly any evidence on "what makes a good interactive book". In this article we provide an in-depth analysis of design issues, and give account of the exploratory evaluation of experience with "Little Rooster", an interactive book designed and implemented at our Lab, first of all for research purposes. We let 7-8 year old kids to "read" the interactive book, and/or a printed, traditional version with the same content. On the basis of analysis of the video recordings of the reading sessions and the interviews, we list our observations on what children found of the visual and sound design, how they used (or not) the interaction facilities for control and what strategy they followed in mixing reading and interacting. We also investigated the effect of the moving, interactive images on understanding and remembering the narrative of the story, and on understanding concepts nowadays usually unfamiliar to children living in a town. We finish the article with discussing experimental methodological issues and summing up design considerations.

**Keywords:** Interactive book, text comprehension, empirical evaluation.

## 1 Introduction

In course of the TERENCE EU7 project [9] we got aware of the rich potentials of the tablet as new medium for delivering reading (and other educational) materials. We started to critically look at existing apps [1] and explore the possibilities by making interactive books [4] in the Creative Technology Lab of MOME [2]. We immediately noticed that the market is running ahead of the sporadic research [4, 6-8]: though the educative apps, among them interactive books are mushrooming, there are hardly any approved design principles, or empirical research on the effect of design decisions. This is no wonder, as the medium offers a richness of possibilities never seen before for creating interactive experiences.

In this article we focus on interactive books for kids (further: IB), designed for tablets. By the term we mean textual reading material which is enhanced with interaction facilities, to "bring to life" the illustrations (which may be additions to some illusion of aliveness). Besides the moving images, features like sound effects, music, text loud reading or even interactive typo or the inclusion of user-created content (drawing,

T. Di Mascio et al. (eds.), *Methodologies and Intelligent Systems for Technology Enhanced Learning*, Advances in Intelligent Systems and Computing 292,
DOI: 10.1007/978-3-319-07698-0_14, © Springer International Publishing Switzerland 2014

photo, own voice recording) may enhance the "reading" experience. Next to (or fully interwoven with) the story, games and puzzles may also be offered, for fun, or to help to understand the text or test what was read or to improve skills.

In this article we give a report on the in-depth evaluation of how 7-8 year old kids used and experienced "Little Rooster". This IB was designed and implemented at our Lab with the intention to use it as a "medical horse" to research how children react to phenomena they do not encounter in traditional books.

In section 2 we introduce the story and the design of "Little Rooster", and the experimental setting in which it was tested. In section 3 we discuss some important results of the evaluation of usage, design, subjective experience and effect on text comprehension. We finish the article with summing up major conclusions and discussing experimental methodological issues, as well as outlining further work.

## 2     The Research Materials and Setting

### 2.1     The Design of "Little Rooster"

We prepared for 6-8 year old kids an interactive version of a well-known Hungarian folk tale. It is about Little Rooster who finds a diamond halfpenny which is confiscated by the Sultan, but finally taken back by the cunning bird. The story develops in 4 scenes, in each the Sultan is trying to torture Little Rooster to stop him from demanding back her halfpenny. But every time Little Rooster outwits the Sultan, and finally gets not only his halfpenny back, but takes all the treasures of the Sultan. In each of the scenes there is some characteristic movement, and objects from peasant households (a well, a bread-baking oven and a bee-hive) play an important role. The text is relatively short and balanced. There are three major conflicts, each involving 2 symmetrical situations, and each conflict is solved by the Little Rooster repeating some similar spells, till the danger is over (sip up all the water, suck up all the bees, let out all the water, let out all the bees).

After having looked at the visual and interaction design of dozens of interactive book titles from the app stores, we decided to create of our own IB, with the flowing goals:

1. The illustrations should be of outstanding quality, original in appearance, and different of the „ruling" vector graphics style.
2. If the depicted phenomena (like fire, or bees, or the hub of a wheel of a well) have sound, the moving images should be enhanced with the proper sound effect.
3. The majority of the „interactives" should underpin the story, its narrative – by letting the child to re-enact the major elements in the narrative. (We will refer to these as narrative interactives, N.)
4. For experimental purposes, we include a few interactives which enrich the scenery. (We will refer to these as enrichment interactives, E.)
5. Beside touch gestures, shake is also to be considered as input modality.
6. The interactions should be designed logical enough that the child can discover them easily, without any visual or textual help.

7. We use all aspects of the design to support the text comprehension of the child, relying on narrative principles [5].

The illustrations were made by a graphic artist specialised in children's book illustrations. We consulted with experts in graphic design, in book typography, but also with a professor in pedagogy specialised in education of young children. The illustrations were drawn by hand, with pencil on paper. They turned out to be so rich in texture that, after some preliminary tests (by kids, as well as by graphic designers and educators) with colourings, we dropped the initial idea of colouring the illustrations, and used them in monochrome version, underlining the archaic style.

For the textual part, we used Foglihten sheriff font (pt 26) which is clean and easy to read on tablets by early readers. For each page the text was carefully placed with respect to the illustration. We enhanced the simple text with some words in red (42 point) or in bold (35 point). These features not only made the text attractive, but helped to grasp the essence of the scene (in bold), and to get the action the reader was supposed to animate in the drawings (in red).

A separate page was devoted for each scene in the story (see Figure 1). What could be "brought to life" by interaction in the illustrations, was slightly moving with natural-looking idle movement, attracting the attention. Usually, there was one N per page, corresponding to the major event told in the story on that page. For each page, navigation arrows were placed in the bottom corners, plus an icon to re-start the story.

**Fig. 1.** Two screenshots from "Little Rooster"

Special is the scene where the rooster, if tapped, sips up the water in the well. Then the reader may pull him out of the well by turning the wheel and lifting the bucket. This second interaction extended the text, where there was nothing said about how the rooster got out of the well. This second interaction was conditional, could be performed only after the first act was accomplished.

For events which have sound in nature, sound effects were added, and emotional outburst (e.g. by the Sultan when stung by bees) were used.

The programming was done in Flash. The IB is available for free both for Android and iOS devices, in English and Hungarian. [3]

For comparison, we also created a printed booklet-like version, where the most characteristic moment of the interactive illustration was shown.

## 2.2     The Setting of the Evaluation

We set out at performing an in-depth empirical analysis of usage, revealing details of interaction and providing ground for methodology for further, large-scale evaluations. We went to a local primary school in Budapest, and had sessions with 17 children, each from class 2 (just learnt to read), aged 7-8. Children were selected to represent a balanced group with respect to gender, reading skill (indicated by the teacher) and media exposure (assessed in a preliminary interview). Children had an interaction session one by one, alone, in a separate room used for after-school activities. The sessions were conducted by well prepared students (who emphasized that they were not the designer). The entire session was video recorded automatically, resulting in a 25-45 minutes video for each session.

The child was told that he/she would be contributing to testing a new kind of book designed for their age group. Then the child was briefed to asses reading habits, familiarity with the story (they were asked to tell the story which served as basis for comparison after the session), and media usage. Then the child followed one of the procedures, depending in which group he/she participated.

In **T (tablet) group**, she/he was introduced in 2 minutes to the touch gestures which appear in Little Rooster (tap, circular) and shaking, by manipulating a scene with a boat (made for the purpose). Then she/he was told to read the Little Rooster, play around with the images, in the way she/he likes, as long as she/he prefers. Then an interview was taken, with questions about usage and liking of the app, the illustrations, the sounds. The story had to be told by the child. The conductor asked to explain some objects or events which came to life by interaction. The latter questions were intended to get an idea about text comprehension.

In **P (print) group**, the child was given the printed flip-over booklet, and similar process was pursued as for T. Then, as a bonus, the child was shown the interactive version, with the same intro as in group T. After the child finished to play around with the IB, also the questions relevant to interaction design were asked. She/he was also asked about preference. Due to space limitations, in this article we use the P group as control group.

Each video recording was coded by 2 coders, independently. In case of discrepancy, the third author took a critical look and resolved the conflict. Besides the interviews, the usage of the app was also coded: for each interaction in each scene the numbers of successful and failed trials were counted, as well as time periods of reading only, interacting only, and both. The coders were also looking for emotional reactions (by face, in words, body).

# 3     Evaluation Results

## 3.1     Interactions

Table 1 gives an overview of the interactions performed by each subject in group T. We sum up observations on **interaction types**, taking also into account additional data from the coding of the usage and interviews:

**Table 1.** Types of interactions performed in group T. Data for E types are in grey rows.

| effect of interaction | code | number | T1 | T2 | T3 | T4 | T5 | T6 | T7 | average |
|---|---|---|---|---|---|---|---|---|---|---|
| LR pecking | N1 | 5 | 4 | 3 | 3 | 5 | 1 | 5 | 2 | 3,29 |
| LR jumps while running | N2.1 | unlimited | 1 | 1 | 2 | 2 | 1 | 0 | 0 | 1,00 |
| S sits down while running | N2.2 | unlimited | 0 | 0 | 3 | 3 | 0 | 6 | 0 | 1,71 |
| LR crowing | N3 | unlimited | 2 | 1 | 2 | 5 | 1 | 2 | 2 | 2,14 |
| LR sipping up the water | N4.1 | 3 | 3 | 3 | 3 | 3 | 2 | 3 | 3 | 2,86 |
| rescue LR by turning the wheel of the well | N4.2 | 1 | 0 | 2 | 1 | 2 | 0 | 2 | 2 | 1,29 |
| bird singing | E1 | unlimited | 0 | 2 | 2 | 0 | 0 | 2 | 2 | 1,14 |
| LR jumping away from servant | N5.1 | unlimited | 2 | 1 | 2 | 6 | 1 | 0 | 0 | 1,71 |
| servant jumping up to catch LR in vane | N5.2 | unlimited | 0 | 0 | 0 | 5 | 4 | 7 | 3 | 2,71 |
| weathervane turning | E2 | unlimited | 0 | 1 | 1 | 0 | 0 | 0 | 2 | 0,57 |
| LR putting out the fire | N6 | 6 | 0 | 6 | 6 | 6 | 3 | 6 | 6 | 4,71 |
| LR sipping up the bees | N7 | 4 | 1 | 4 | 4 | 4 | 4 | 4 | 4 | 3,57 |
| single bee flying away from flower | E3 | unlimited | 0 | 1 | 3 | 3 | 5 | 0 | 2 | 2 |
| S suffering from bees | N8 | unlimited | 0 | 1 | 3 | 7 | 4 | 9 | 1 | 3,57 |
| LR sipping up coins | N9 | 3 | 3 | 3 | 3 | 3 | 3 | 3 | 3 | 3 |
| waggle rocking chair | E4 | unlimited | 0 | 0 | 0 | 0 | 0 | 0 | 0 | 0 |
| tablet in hand | -- | --- | Y | Y | Y | Y | Y | N | Y | 6/7 |

1. N interactions were preferred over E interactions, that is, children indeed were "replaying" the story.

2. Children preferred more interactions which tortured the Sultan (S). This is particularly apparent where pairs of N interactions could be done in the same scene, one of them punishing the S (or his servant). Children clearly enjoyed the most the nearly sadistic interaction N8 (this was also expressed in the interviews).

3. Children performed N interactions as many times as needed for completing the event mentioned in the narrative where the goal was clear (N4.1, N6, N7, N9). Interestingly, only less than half of the subjects completed the very first N interaction – finding the diamond halfpenny. This may have to do with the fact that pecking is not so attractive act, and had to be repeated 4 times before success. Unlike in the other cases, here no "partial result" could be seen after each tap.

4. No child performed E4, though they usually held the tablet in hand. This was the only E interaction by "shake" gesture. Moreover, this was in the closing "happy end" scene of the story.

5. The specific circular gesture (to turn the wheel of the well) was not properly by most the subjects. This interaction was not suggested by the story, but – just as all the other ones – indicated by slight idle movement. As the children were not familiar with how the depicted well works (this was evident from the interviews

too), so the lack of real-life example prevented them from performing the right gesture.

6. There were characteristic differences in subjects' attitude. E.g. T1 was very reluctant to explore the images, while T6 was the most active. (Both subjects were familiar with tablets, T1 was girl, T6 was boy).

7. Navigation was no problem for any of the subjects (not shown in the table).

As of **reading strategy**, here are some of our observations:

1. All children did read the text, all but one fully, often even second time (in group P, or when repeating the page).

2. Majority of the children read the text first, and then started to explore the illustrations. A couple of subjects were jumping between reading and interaction. Only 2 subjects out of the 17 started to explore the illustrations before reading.

We can say that neither the interactive content nor the sound effects distracted the subjects from the reading. This may have to do with the school setting, but may be also the positive effect of the typo with eye-catching features.

## 3.2     Design

As of the **illustrations**, we found:

1. Children did not complain about the monochrome and archaic style. On direct question, only a few said that they would prefer to have some colour. Nobody mentioned "vector graphics style" as more preferred or appropriate on tablets.

2. Children were very critical about details of the drawings. Several boys were asking for (yet) more realism in the "living" scenes.

3. Children were pinpointing small discrepancies between story and the effect of interaction.

Children did notice the **special typography**. As of the function of the highlighted words, we can say:

1. Highlighted words drew their attention, some kids tapped on them at first.

2. In the interviews about a third of the subjects attributed to these highlighted words the "essence" of the story. Some discovered their "hint" role for interactions.

Feedback about the **sounds** was very sporadic. Here are the general conclusions:

1. The permanent and dramatic "background sound" (in one scene the noise of continuous chasing and running) did disturb poor readers – this could be noticed in their reading behavior as well as what they told in the reflective interview.

2. The sound of the natural phenomena (bees, fire, …) did not occur to them. On the other hand, in recalling the story some kids did use the special sound of the rooster crawl. (Our questioning was not addressing the sound effects per se.)

### 3.3 Liking and Text Comprehension

We explored **liking** on the basis of questions in the interviews. Here are the major observations:

1. Most of the subjects said they liked the interactive tale. They referred to the moving image, possibility to be involved in the story. Several children mentioned the joy of punishing the Sultan. "Funny" was a common term used to explain liking.
2. Several children, especially boys said that this tale is too childish for them, so they would recommend it to younger children. (Actually the tale is told in kindergarten, and many children knew it from years before, though they forgot the events.) Boys also expressed that they would prefer to have more action, more excitement. There was a clear gender difference in liking the story, even in the IB form.
3. All children were open for another similar experience.
4. The (facial and bodily) behavior of most of the children did not reveal liking – they were "going through" the IB as a (school) task, with tense expression.
5. All of the subjects in group P found the IB version more interesting, they would prefer to read a tale in this format.

We were the most interested in the effect of the IB on reading and **text comprehension**. Here we raise only the major observations based on pre- and post-test inter-views:

1. For T6, a poor reader but experienced in media, this form of reading was clearly much more motivating than for the average. He got most excited, had the most comments on the realization. Besides doing the most interactions (as shown in Table 1), he was reading with interest.
2. We found a correlation between how good reader a child is and how handy she/he is with the interactive media. This may be characteristic of the social conditions at the school, accommodating children of well to do and above average educated parents. Moreover, when looking at the effect of the IB in terms of text comprehension, we found that if the child was a good reader, or experienced with interactive media, he/she was good in recalling what he "read" in the IB. This is in line with the above observation.
3. As of learning new concepts, the "village well", depicted (only) in the interactive illustration, brought results contrary to our expectations. Not only children could not "operate" the well with the required circular gesture, due to lacking the example from real life, but those who had some idea about a (different) well got somehow confused that the well depicted was not like the one they new of. Thus it seems that the knowledge gained from real life makes children richly interact with (realistic) moving illustrations. We cannot (and should not) expect too much in the other direction, at least in this genre and with this age group.
4. The recall of the narrative of the story (number and order of events), as well as usage of special (archaic) terms in the text increased clearly in both the T and P

groups, in comparison to how the story was recalled at the beginning of the session. There seemed to be a better performance in the T group.

## 4    Discussion

In this paper we gave a summary of the exploratory empirical testing of an interactive book by 7-8 year old children. Our most important messages are:

1. Our age group was (still) open to non-commercial, artistic visual representation, and accepted well monochrome drawing on tablets. However, they were critical on content, especially if something did not match their own concept or was not fully in line with what was read in the text.
2. Children "used" different interaction options according to their role in the narrative, and their own sympathy for the characters involved.
3. The interactive medium can make reading appealing, even for poor readers.

The detailed comparison of the T and P groups will be done in another article. Based on the present study, we intend to refine our research setting and target group. We plan the following further investigations:

1. Repeating the experiment with English target group, who are not familiar with the story, and have a different cultural background.
2. Making similar investigations with pre-school children (text narrated).
3. Repeating the experiment with children in a relaxed, home setting.
4. Testing the "vocabulary forming" potential of interactive moving images with objects and events totally unknown to the target group.

**Acknowledgement.** The research was partly done in the framework of the EU FP7 TERENCE and of the "Interactive Book Research" supported by the Chief Architecture Office of the Ministry of Internal Affairs. The Little Rooster was illustrated by Barbara Szepesi-Szűcs, animated and programmed by Dániel Karasz. We thank Dóra Popella, Monika Gyúró and the pupils and teachers of the Budenz School in Budapest for their cooperation.

## References

1. Eaton. K: Tracing Images and Heeding Voices to Learn the Basics of Reading. The New York Times (May 29, 2013)
2. IB projects of the Creative Technology Lab see, http://create.mome.hu
3. Little Rooster is available in English (and in Hungarian) on Google Play, for download info see, http://techlab.mome.hu/felkrajcar
4. Morgan, H.: Multimodal Children's E-Books Help Young Learners in Reading. Early Childhood Education Journal, 1–7 (2013)
5. Murray, J.H.: Hamlet on the Holodeck: The Future of Narrative in Cyberspace. The Free Press, New York (1997)

6. Roxby, P.: Does technology hinder or help toddlers' learning? BBC News (April 19, 2013), http://www.bbc.co.uk/news/health-22219881
7. Plowman, L., McPake, J.: Seven myths about young children and technology. Childhood Education 89(1), 27–33 (2013)
8. Plowman, L., McPake, J., Stephen, C.: Extending opportunities for learning: the role of digital media in early education. In: Suggate, S., Reese, E. (eds.) Contemporary Debates in Child Development and Education, pp. 95–104. Routledge, Abingdon (2012)
9. TERENCE project web site, http://www.terenceproject.eu/

# Opensource Gamification of a Computer Science Lecture to Humanities Students

Giovanni De Gasperis and Niva Florio

Dept. of Information Engineering, Computer Science and Mathematics,
University of L'Aquila, Via Vetoio, Coppito, L'Aquila
{giovanni.degasperis,niva.florio}@univaq.it

**Abstract.** Opensource software can provide a wide range of educational computer games and gaming tools so to enable a sharing community among teachers. Since their appearance Metaverses[1] appealed the technology prone teachers, increasing the number of virtual worlds learning projects. The Open Source movement create an ecosystem of software, documentation and tools that it makes easier for teachers to share their work and learning objects. Here we show how an open source Metaverses becomes a tool for the teaching of science subjects in a Department of Human Studies, where students can learn the basics of a simplified von Neumann-class computer architecture in a OpenSimulator virtual learning environment. A survey among students showed that interacting with a virtual computer architecture in a immersive 3D environment increased their learning achievements and subject comprehension.

**Keywords:** gamification, game learning for TEL, TEL case studies, computer architectures.

## 1 Introduction

Since the appearance of personal computers, educators have been inspired by the use of computer-based technology in learning [12]. More Specifically, referring to pedagogical theories that see playing games as a form of learning, computer games can be used with educational purposes in a technology enhanced learning context [9]. On the other hand, Internet contributes to increase the number and variety of tools that can be used for technology enhanced learning and seems to be a suitable place for the development and playing of educational computer

---

[1] In 1992 Neal Stephenson used for the first time the term metaverse in his novel *Snow Crash* [16]. In the novel it is described as a virtual reality shared on the fiber-optic worldwide network populated by avatars and software agents. Avatars are interactive 3D models of human beings in a virtual reality environment, while a software agent is a software which acts with some autonomy to accomplish tasks in place of the human user belongs to. The metaverse in the novel is then represented as a virtual world where human users in the form of avatars interact with each other and with software agents in an immersive three-dimensional space that uses the metaphors of the real world.

T. Di Mascio et al. (eds.), *Methodologies and Intelligent Systems for Technology Enhanced Learning*, Advances in Intelligent Systems and Computing 292,
DOI: 10.1007/978-3-319-07698-0_15, © Springer International Publishing Switzerland 2014

games; for example by providing students with 3D virtual environments, they can **learn-by-doing** in situations that are difficult or impossible to reproduce in reality (e.g. [5] [10]). As well as computer games, Metaversess inspired teachers since their first appearance, after the personal computer introduction in 1981. For example, tools such as Habitat [14] appeared in 1986; the idea was to give students an imaginative virtual world where to freely experiment socially new ways of learning, inspired to constructivism and collaboration. Also in [7], after analyzing the different types of constructivism and how the interaction with the environment is necessary to build knowledge, the authors demonstrate how virtual worlds used with educational purposes fall within the educational theory of constructivism. Also in [11], referring to the definition of Piaget's constructivism, the authors show how 3D virtual worlds facilitate the activities of constructivist learning because students can experience and have the opportunity to learn by doing. Metaversess offer new opportunities for the acquisition of knowledge related to the action and to the interaction with the environment [13] [18]. Learning through Metaversess by means of student avatar is bound to discover and to do themselves, to participate actively in the creation and development of their own knowledge [13] [18]. Metaversess are the ideal place to build scenarios related to learning through problem-solving and hands-on training activities for exploiting into practice learners knowledge [19] [6].

Metaversess are the ideal place to create simulation and/or role games, even when adopted in abstract scientific contexts. Referring to the definition of "game" in [9], we present a simulation game in which there are no scores to reach or levels to achieve. We tried to give students just the opportunity to experience the interaction with an object that is difficult to replicate in real life, in order to improve the understanding of the abstract von Neumann machine operations. In this way we gamified the abstract concept of a computer architecture.

In Section 2 some examples of educational application of Metaversess in different fields are presented; then in Section 3 we describe how an opensource Metaverses become an educational tool for the teaching of science subjects in the Department of Humanities of the University of L'Aquila; in Section 4 we conclude.

## 2    Learning in Virtual Worlds

Virtual worlds are a great help in learning projects among different disciplines [1], from scientific ones to liberals arts and human sciences, as the following non exaustive list of use cases can show.

MetaRezzer [3] is an experiment in teaching basic computer science, information processing systems and fundamentals of procedural, event-driven and object-oriented programming languages. It is a web-Metaverses that allows to define objects manipulated in a three dimensional environment via a web interface, which interacts with the server to generate the Metaverses objects defined by the user.

In [17] a teaching and learning platform is adopted to improve learning in computer science for students of the Computer Information System Department at Borough of Manhattan Community College. As in real worlds students have a lecture area where they can attend lessons on relevant topics using a shared virtual slide projector and a teacher avatar, and then participate in virtual lab activities. These activities have been developed for a selection of topics in computer programming (loop, array, etc.) and they are 2D simulation with practice questions, 3D simulation and 3D games. On this platform students have also study rooms where they can discuss and collaborate for study purposes. The Engineering Education Island [2] is realized to investigate how virtual worlds can improve learning in engineering. It was developed a virtual laboratory where students can interact with demonstrations and simulations, a lecture hall where they can attend classes and other facilities where they can work together. In particular, the authors create a demo to help students understanding how the CPU cycle works: they realized the various component of the cycle and reproduce it in different steps to illustrate how CPU perform single arithmetic operations on different instruction data.

English Community is a virtual learning environment for Aijia Cambridge English Training Center [8]. It is a 3D virtual environment for project-based learning approach, where students are encouraged to communicate in English, assuming that it is difficult for students to immerse themselves in cultural and social reality of a foreign country if you do not visit it. This virtual world consists of an activity center, restaurants, cafes, lecture halls, a post office, a supermarket and a virtual gym, where students can do spare time activities, follow lessons, meet to discuss about a topic and to do group learning activities. In a typical learning session, the teacher asks a student to go shopping at the supermarket, the student has to complete the activity learning how to get what s/he needs in the supermarket by communicating with others.

In [10] the Best Digital Village is described. In this 3D virtual learning platform, through role playing games and multi-learner engagement, students become familiar with the parabola flight-formula and other related tasks. With this village, Kuo and Lin intend to propose not only scientific contents, but also human contents, by means of social learning process and teamwork gaming. For example, using the parabola flight-formula, team members works together to realize a virtual catapult to hit a target in the park of the village. Before playing this game, students have to complete individually and successfully other learning tasks to make the team qualified for the catapult game.

The Laconia Acropolis Virtual Archeology (LAVA) [5] reproduces a site of archaeological interest which exists in the real world, a Byzantine basilica in the Sparta region in Greece; they can explore at firsthand the archeological site, starting from the concepts learned in the classroom. This kind of scenario avoids problems which in the reality are common for archeology students: the high financial costs of participating to a real excavation project, the limited number of students permitted for an archeological site, the physical distance of the more interesting sites, the non-recoverability of a mistake. With LAVA students virtually live all the steps of an excavation, from the submission of the proposal to the exhibition of their findings, going through the different steps of the simulation and meeting a series of domain-specific objectives.

These non exhaustive list of use cases of Metaverses in learning projects have in common that target students usually **share the same background** of teaching disciplines, i.e. teaching engineering related disciplines to engineering students. Appears to be a lack of experiences of virtual worlds learning projects where target students have a completely different study curriculum from the discipline, i.e. learning computer science basics for liberal art and humanities students.

# 3   Case Study

The case study of a collaborative learning setup in Metaverses is a simulator for a virtual von Neumann-class machine. The subject is important to let students understand the actual behind-the-curtain reality of a computing machine, also to became well educated computer-aware users. Over the years we have realized that this topic is particularly stubborn and difficult to understand for students with a curriculum of humanities. For this reason, in a private OpenSim land of the Faculty of Humanities of the University of L'Aquila [4], a simplified version of the standard von Neumann machine has been built and scripted, in order to help students better understand this type of architecture and its running cycles. In a typical lesson, students interact with this simulator and, through virtual touching and chat interactions, they can activate different behaviors of the machine, such as reading and writing data from memory or reading and writing data through virtual peripherals.

## 3.1   Simplified von Neumann Machine

The architecture of most modern computers is organized according to the architecture designed by the John von Neumann in the 1940's [15], which consists of four basic functional elements: i) the **central processing unit** (CPU), containing the digital electronic circuits capable of acquiring, interpreting and executing the instructions of a program, i.e. registers, arithmethic logic unit, accumulator, micro-code interpreter; ii) the **main memory**, the device where the necessary information to execute a program (instructions and data) are stored; iii) **input and output devices**, to transfer information between main memory and/or CPU and the external environment; and iv) the **system bus**, which links the various components, trasfering memory addresses and data . The processing unit coordinates the various activities: it extracts instructions from memory, decodes them as microprogrammed in the hardware and runs them. The content of memory is routed based on location and the operation address, regardless of the type, i.e. data or instruction, and the instructions are executed in strict sequential order. This process constitues the machine cycle (CPU cycles or Fetch-Decode-Execute loop), which can ideally be divided into three phases: i) the **fetch** phase when the control unit retrieves the instruction from the main memory and increments the value of the Program Counter to point to the next instruction; ii) the

**Fig. 1.** Simplified von Neumann virtual machine. Top is the ADDRESS BUS, bottom is the DATA BUS. From left to right: memory, with internal memory locations (at different heights so the address can be counted), CPU, Input and Output devices. Inside the CPU, ALU and some registers are visible. The student gives the command *"loc01 write 45"* and the virtual machine simulator responds with a colored animation that visualize the storage of the value 45 in the memory cell loc01. In the chat window are shown the messages shown following a general *"memory reset"* command.

**decode** phase when the instruction statement is micro-interpreted; iii) in the **execute** phase when the CPU sends appropriate signals to the hardware that implement the commands for the execution. The machine cycle is marked by a timer or a clock: an oscillator that emits signals at regular time intervals, which plays a primary engine of operation. We adopted the splitted bus solution into ADDRESS and DATA bus, as usually done in most recent von Neumann-class architectures, also because it helps to clarify data flow to first time students. The overall architecture implemented with LSL scripts in OpenSim is shown in Fig. 1, with an example of interaction with student-avatar.

### 3.2   Students Interaction

Students, driving their avatars, can interact among themselves and with the virtual von Neumann-class machine giving commands through the public chat. To write a value in a single memory location they could write the command *loc01 write 45*, the virtual machine will behave highlighting the complete data flow arriving at the selected memory location, giving it a color visualizing the value it has inside, like shown in Fig. 1.

If students gives a more complex command, like *cpu write 1 12*, it generates a cascade of commands from the CPU object such as the following:

$$> \ Jodeg\ Janus: \ cpu\ write\ 1\ 67$$
$$CPU: \ ABus\ 1$$
$$CPU: \ DBus\ 67$$
$$ABus: \ MEMORY\ ADDRESS\ 1 \tag{1}$$
$$DBus: \ MEMORY\ DATA\ 67$$
$$CPU: \ Memory\ WRITE$$
$$Memory: \ loc01\ write\ 67$$

The process is properly visualized during execution of the commands highlighting the components that are involved in the data-flow. The implemented functionality are: memory read, memory write, output a value, input a value. Also basic instructions could be saved in the memory simulating the fecth, decode, execute, write back of the full CPU cycle. The overall learning cycle experience can be summirized as the following: write the code and see what happen in the trasparent lightned and color animated architecture, which make the fun out of it, reoinforcing the student learning if the outcoming behaviour is as expected. Also students avatars can discuss what the virtual machine is doing by chat or by voice, if in the same classroom, and live a collaborative learning experience with mutual reinforcement and exchange of ideas.

**Table 1.** Sample: 15 humanities undergraduate students, age: 19-25. Method: Moodle anonymous online polling taken after a week from the lecture in the classroom.

| Question | Yes % | No % |
|---|---|---|
| Before the course I did not know the concept of Metaverses | 93.33% | 6.67% |
| Now I know and understand what is a Metaverses | 93.33% | 6.67% |
| I attended the introductory lecture on Metaversess | 86.67% | 13.33% |
| I have an avatar in a Metaverses | 13.33% | 86.67% |
| I already had experience of building virtual objects | 20.00% | 80.00% |
| Now I can create objects in a virtual Metaverses | 73.34% | 20.00% |
| I attended the lecture about the construction in OpenSim and interaction with the von Neumann Virtual Machine | 80.00% | 13.33% |
| I attended all the lectures on architecture and computational models (Turing machine, Von Neumann machine) | 86.67% | 13.33% |
| I had already understood the operation of the architectures and computational models from theoretical lessons | 46.67% | 46.67% |
| The operation of the von Neumann machine was already clear to me from the theoretical lessons | 40.00% | 40.00% |
| Before the exercise with the virtual von Neumann machine did not understand its operation | 40.00% | 26.67% |
| After exercise with the virtual von Neumann machine I figured out how it works | 46.67% | 20.00% |
| Exercise with the virtual von Neumann machine was helpful to clarify the doubts I had about theoretical lessons | 73.33% | 0% |
| Lessons and exercises in the Metaversess did not add anything to my learning process | 6.67% | 73.33% |
| After exercises on virtual world I have confused ideas on the functioning of von Neumann machiene theory that I thought I had understood | 0% | 80.00% |
| I wish I could do other exercises in virtual reality on other topics of Computer Science | 53.33% | 20.00% |

### 3.3 Students Feedback

The 15 humanities students involved in the Metaverses based lecture were asked to give feedback about their learning experience after using the gamefied virtual von Neumann-class machine. The students participated in an anonymous survey with questions about the learning event. As shown in Table 1, the majority of students says that the von Neumann virtual machine helped them to understand this architecture and its operations, and they found it so useful that they would like to do this kind of exercises also on other topics. Since the introduction of this kind the lecture has been observed a general qualitative improvement at the final exam over the years.

## 4   Conclusion

Here we proposed to adopt the Metaverses as a immersive computer game with learning purposes in order to advance the knowledge of technical disciplines in students coming from a non technical background such as students with arts and humanities curricula. The immersive virtual environment helps student to visualize abstract mathematical and technological concepts and gives them the basis to fully understand what is needed for a basic computer science course for higher education. In the future a full simplified machine language instruction set, including jump instructions, could be implemented letting students of a more advanced programming course to get familiarized with the Assembly language.

## References

1. Bainbridge, W.S.: The Scientific research Potential of Virtual Worlds. Science 317(5837), 472–476 (2007)
2. Callaghan, M.J., McCusker, K., Losada, J.L., Harkin, J.G., Wilson, S.: Teaching Engineering Education Using Virtual Worlds and Virtual Learning Environments. In: International Conference on Advances in Computing, Control, and Telecommunication Technologies (ACT 2009), pp. 295–299 (2009)
3. De Gasperis, G., Salvi, A.: Uso e strumenti basati su Metaversi 3D per l'insegnamento dell'Informatica e dei Sistemi di Elaborazione delle Informazioni. In: Proceeding of DIDAMATICA 2009 (2009)
4. De Gasperis, G., Di Maio, L., Di Mascio, T., Florio, N.: Il metaverso Open Source: Strumento didattico per facoltà umanistiche. In: Proceedings of DIDAMATICA 2011 (2011)
5. Getchell, K., Miller, A., Nicoll, R., Sweetman, R., Allison, C.: Games Methodologies and Immersive Environments for Virtual Fieldwork. IEEE Transactions on Learning Technologies 3(4), 281–293 (2010)
6. Gorrino, A., De Gasperis, G.: Virtual laboratory for the training of health workers in italy. In: Omatu, S., Paz Santana, J.F., González, S.R., Molina, J.M., Bernardos, A.M., Rodríguez, J.M.C. (eds.) Distributed Computing and Artificial Intelligence. AISC, vol. 151, pp. 41–48. Springer, Heidelberg (2012)
7. Girvan, C., Savage, T.: Identifying an appropriate pedagogy for virtual worlds: A Communal Constructivism case study. Computers & Education 55, 342–349 (2010)

8. Jiang, X., Liu, C., Chen, L.: Implementation of a project-based 3D virtual learning environment for English language learning. In: 2nd International Conference on Education Technology and Computer (ICETC), vol. 3, pp. 281–284 (2010)
9. Jong, M.S., Lee, J.H., Shang, J.: Educational Use of Computer Games: Where We Are, and What's Next. In: Reshaping Learning, pp. 299–320. Springer, Heidelberg (2013)
10. Kuo, M.S., Lin, C.S.: Virtual Parabola Festival: The Platform Design and Learning Strategies for Virtual Learning Community of Practice. In: 3rd IEEE International Conference on Digital Game and Intelligent Toy Enhanced Learning (DIGITEL), pp. 3–9 (2010)
11. Leman, F.G., Gu, N., Williams, A.: Virtual worlds as a constructivist learning platform: evaluations of 3D virtual worlds on design teaching and learning. IT-con. Special Issue Virtual and Augmented Reality in Design and Construction 13, 578–593 (2008), http://www.itcon.org/2008/36
12. Liu, M., Moore, Z., Graham, L., Lee, S.: A look at the research on the computer-based technology use in second language learning: a review of the literature from 1990-2000. Journal of Research on Technology in Education 34(3) (2002)
13. Morganti, F., Riva, G.: Conoscenza, comunicazione e tecnologia. LED Edizioni Universitarie, Italy (2006)
14. Morningstar, C., Farmer, F.R.: The Lessons of Lucasfilm's Habitat. Paper presented at the 1st International Conference on Cyberspace at the University of Texas, Austin (1991)
15. Von Neumann Architecture, Wikipedia, http://en.wikipedia.org/wiki/Von_Neumann_architecture (accessed September 9, 2013)
16. Stephenson, N.: Snow crash. Bantam Books (2000)
17. Wei, C.S., Chen, Y., Doong, J.G.: A 3D Virtual World Teaching and Learning Platform for Computer Science Courses in Second Life. In: International Conference on Computational Intelligence and Software Engineering (CiSE 2009), pp. 1–4 (2009)
18. Zhao, H., Sun, B., Wu, H., Hu, X.: Study on building a 3D interactive virtual learning environment based on OpenSim platform. In: International Conference on Audio Language and Image Processing, pp. 1407–1411 (2010)
19. Zhou, Z., Jin, X.L., Vogel, D., Guo, X., Chen, X.: Individual Motivations for Using Social Virtual Worlds: An Exploratory Investigation in Second Life. In: 43rd Hawaii International Conference on System Sciences, pp. 1–10 (2010)

# Pictorial Representations of Achievement Emotions: Preliminary Data with Children and Adults

Daniela Raccanello and Caterina Bianchetti

University of Verona, Italy
daniela.raccanello@univr.it, caterina.bianchetti@gmail.com

**Abstract.** Three studies are presented aimed at testing a preliminary version of a pictorial instrument representing children's achievement emotions as conceptualized in control-value theory. Children (second- and fifth-graders) and adults were administered three tasks assessing the correspondence between drawings of faces and ten achievement emotions (enjoyment, pride, hope, relief, relaxation, anxiety, anger, shame, boredom, and hopelessness): an agreement task (Study 1, n = 46), a matching task (Study 2, n = 47), and a naming task (Study 3, n = 53). Analyses on the agreement and matching task revealed accurate responses for all the emotions, while in the naming task low accuracy emerged for pride, hope, relief, and particularly for boredom. Results are discussed in light of their applicative relevance for the future development of the instrument.

**Keywords:** Emotional Faces, Drawings, Achievement Emotions, Children.

## 1    Introduction

Despite the long recognized relevance of emotions associated with learning, they have only recently assumed a crucial role within educational psychology and are increasingly studied according to their mutual relationships with cognitive, motivational, and behavioral dimensions [17, 20]. Concerning learning contexts, emotions – as multi-componential phenomena "including affective, cognitive, motivational, expressive, and peripheral physiological processes" [19, p. 316] – have historically been studied as related to causal attribution processes [26] and test anxiety [28], and only in the last decade has attention been paid to achievement emotions, rather than to states pertaining to different valences [21]. According to theoretical frameworks such as control-value theory, achievement emotions can be defined as those emotions related to achievement activities or outcomes, and they can be conceptualized considering two underlying dimensions, valence (positive versus negative emotions) and activation (activating versus deactivating emotions) [19]. This model, based on assumptions from different theoretical perspectives such as expectancy-value theories of emotions, transactional approaches, and attributional theories [15, 25, 26], considers achievement emotions in terms of their antecedents and outcomes. Feelings of control – or lack of it – like self-efficacy beliefs and

T. Di Mascio et al. (eds.), *Methodologies and Intelligent Systems for Technology
Enhanced Learning*, Advances in Intelligent Systems and Computing 292,
DOI: 10.1007/978-3-319-07698-0_16, © Springer International Publishing Switzerland 2014

task-value are assumed as the proximal antecedents of emotions. In turn, achievement emotions would deeply influence learning and performance, also by mediating mechanisms such as motivation, strategy use, and learning regulation.

Achievement emotions can be assessed by means of questionnaires such as the Achievement Emotions Questionnaire (AEQ), focused on nine emotions and developed for university students [20], or the Achievement Emotions Questionnaire-Elementary School (AEQ-ES), which measures elementary school students' enjoyment, anxiety, and boredom, also via graphical displays [16]. However, there is a lack of measurement instruments anchoring a wide range of achievement emotions to the corresponding pictorial representations, which would favor their evaluation with children or individuals with special needs. Such instruments would be particularly relevant from an applicative point of view in the design of accessible, usable, and effective Technology Enhanced Learning (TEL) products aiming at assessing emotions characterizing learning contexts.

With the wider diffusion of new technologies, affective facial stimuli such as photographs or drawings have increasingly been used in emotion research. Many studies have focused on the recognition of basic emotions such as happiness, sadness, anger, and fear in adults and children, also with atypical development like autism, schizophrenia or intellectual disabilities [e.g., 4, 11, 14, 18, 24, 27]. To avoid methodological problems that could affect reliability and validity of findings, some researchers have proposed standardized sets of emotional stimuli like photographs, in some cases specifically intended for children [e.g., 6, 11]. However, less attention has been paid to standardized sets of drawings, particularly familiar to children and also characterized by higher ecological validity compared to schematic or computerized faces [e.g., 10, 23]. Processing drawings representing emotions (whether associated to words or not) instead of only words representing emotions should favor a more direct access to the semantic network in which emotional information is stored [11]. Drawings should consequently be privileged in case, for example, of time constraints typical of some learning tasks associated to computer-based environments.

Recognition of facial emotions has clear links with social functioning: it is a critical element for experiencing emotions and it has a main role in adapting to the environment and facilitating social interactions [7]. Children can already recognize facial expressions of basic emotions at two/three years [3], and their ability gradually develops during the preschool years [5]. Representing knowledge about emotional expressions through a verbal label is more complex than the simple recognition of facial expressions, but also this competence is acquired early, and it is accurately mastered by the age of five, at least for basic emotions [1]. However, matching tasks are solved correctly earlier than naming tasks [2].

Based on existing knowledge, the three studies presented here were conducted with the aim of testing a preliminary version of a pictorial instrument representing children's achievement emotions as conceptualized in the control-value theory of emotions [19]. Study 1 assessed whether and how much children and adults agreed on the correspondence between the pictorial representations of emotions and the proposed emotional labels; Study 2 assessed whether they accurately matched the faces with the corresponding labels; Study 3 assessed whether they accurately named the emotions expressed by the faces. Because of the nature of the tasks, the naming task was hypothesized as more difficult than the matching task, and the matching task more difficult than the agreement task. Greater difficulties were anticipated for achievement emotions different from basic emotions. As a secondary aim, we checked whether children and adults differed in their answers in the three tasks.

## 2    Study 1

Through an agreement task this study aimed at investigating whether and how much children and adults agreed on the correspondence between ten pictorial representations of achievement emotions and the hypothesized labels.

### 2.1    Method

**Participants.** The participants were 46 students from Northern Italy (63% female): 13 second-graders ($M_{age}$ = 8;0; range: 7;6-9;4), 13 fifth-graders ($M_{age}$ = 10;11; range: 10;5-11;10), and 20 university students ($M_{age}$ = 22;7; range: 19;7-38;2), with different social backgrounds. Both in this and in the following studies, adults gave their consent to participate, while written parental consent was obtained for the children.

**Materials and Procedure.** The participants were administered a written agreement task, adapted from the literature [6]. They were proposed ten children's faces representing ten achievement emotions, including three positive-activating (enjoyment, pride, hope), two positive-deactivating (relief, relaxation), three negative-activating (anxiety, anger, shame), and two negative-deactivating (boredom, hopelessness) emotions [13]. The pictorial representations were drawn by a professional illustrator and presented to the users on A4 sheets. In drawing the faces, the minimal elements identified by Ekman and Friesen [8, 9] for representing basic emotions were considered in terms of changes of eyebrows, eyelids, cheeks, nose, and lips. For achievement emotions different from basic emotions, the pictorial representations were based on a combination of such elements, as suggested by Ekman and Friesen [8, 9]. For each emotion, elements representing hair and brows helped elaborate a male and a female version of each face, to favor children's identification. Order of drawing presentation was randomized and kept constant. The participants were asked to evaluate how much they agreed on the correspondence between each face and the hypothesized linguistic emotional label on a 5-point Likert-type scale (1 = no agreement, 5 = very high agreement).

In all three studies, the children were tested together in their classroom, while the adults were tested individually. For each task, both children and adults were first shown one or more faces referred to emotions not included in the ten achievement emotions. They were then asked to respond to the stimuli in the same way proposed for the target emotions in order to familiarize with the task.

## 2.2    Results

For each emotion, agreement on the correspondence between faces and labels was calculated in terms of medians. For pride, hope, relief, relaxation, shame, and hopelessness the median value was 4 (high agreement), while for enjoyment, anxiety, anger, and boredom it was 5 (very high agreement).

In addition, we checked for differences due to age. Kruskal-Wallis tests (level of significance: $p < .01$, in all three studies) revealed that the participants' answers differed according to age for pride ($X^2(2, 45) = 12.880$, $p = .002$) and boredom ($X^2(2, 45) = 10.758$, $p = .005$). For both emotions, the median was 5 for the children, while it was 4 for the adults.

## 2.3    Discussion

These results suggest that the participants evaluated all the drawings as corresponding to the given linguistic emotional labels, with only slight differences related to age.

# 3    Study 2

The aim was to assess through a matching task whether children and adults were able to identify the correspondence between emotional linguistic labels and the hypothesized faces.

## 3.1    Method

**Participants.** The participants were 47 students from Northern Italy (51% female): 13 second-graders ($M_{age} = 7;10$; range: 7;5-8;3), 14 fifth-graders ($M_{age} = 10;10$; range: 10;6-11;3), and 20 university students ($M_{age} = 22;6$; range: 19;7-24;2), with different social backgrounds.

**Materials and Procedure.** The participants were administered a written matching task [12]. They were asked to match the names of ten achievement emotions to the same ten pictorial representations of faces used in Study 1. Positive and negative emotion faces were presented separately on two different sheets, given the need to keep the task as simple as possible, especially for the younger children. For each valence, the order of the faces was randomized and kept constant. Positive and negative labels were written separately at the top of each sheet, and the participants had to match them with the drawings by writing their number next to the corresponding face. For each emotion, answers were coded as accurate (1), when the participants matched the faces to the corresponding labels, or inaccurate/missing (0).

## 3.2    Results

Chi square tests revealed that the percentage of participants who accurately matched the faces to the corresponding achievement emotions was significantly higher than the chance level for most emotions, except for boredom, for which all the participants gave the accurate response (see Table 1 for frequencies, percentages, Chi square tests, and significance levels).

Moreover, Chi square tests carried out to explore differences due to age, separately for each emotion, resulted non-significant.

**Table 1.** Frequency (percentage) of accurately matched responses, Chi square tests (degrees of freedom, $df$), and significance levels ($p$) for each achievement emotion

| Achievement emotions | Frequency (%) | $X^2$ ($df$) | $p$ |
|---|---|---|---|
| Enjoyment | 42 (89%) | 29.128 (1, 47) | < .001 |
| Pride | 37 (79%) | 15.511 (1, 47) | < .001 |
| Hope | 38 (81%) | 17.894 (1, 47) | < .001 |
| Relief | 41 (87%) | 26.064 (1, 47) | < .001 |
| Relaxation | 39 (83%) | 20.447 (1, 47) | < .001 |
| Anxiety | 40 (85%) | 23.170 (1, 47) | < .001 |
| Anger | 46 (98%) | 43.085 (1, 47) | < .001 |
| Shame | 44 (94%) | 35.766 (1, 47) | < .001 |
| Boredom | 47 (100%) | - | - |
| Hopelessness | 45 (96%) | 39.340 (1, 47) | < .001 |

### 3.3    Discussion

The present study indicated that most children and adults, without differences due to age, accurately matched the faces to their corresponding achievement emotions.

## 4    Study 3

In a naming task, both children and adults were asked to give a label for each of the ten achievement emotions represented via faces.

### 4.1    Method

**Participants.** The participants were 53 students from Northern Italy (41% female): 14 second-graders ($M_{age}$ = 7;9; range: 7;5-8;3), 19 fifth-graders ($M_{age}$ = 10;11; range: 10;5-11;3), and 20 university students ($M_{age}$ = 22;10; range: 20;4-27;3), with different social backgrounds.

**Materials and Procedure.** The participants were administered a written naming task [12]. They were asked to write the name of the emotions expressed by each of the ten faces used in Studies 1 and 2. The order of the faces was randomized and kept constant. For each emotion, answers were coded as accurate (1), when the participants named the face accurately, or inaccurate/missing (0). Synonyms of achievement emotion terms were also coded as accurate. A first judge coded all the data, while a second judge coded 30% of the data. Mean percentage of agreement was 90%.

## 4.2    Results

Chi square tests indicated that the percentage of participants who wrote the accurate term was significantly higher than the chance level for enjoyment, relaxation, anxiety, anger, shame, and hopelessness. However, the percentage did not differ from the chance level for pride, hope, and relief, while most participants did not give an accurate linguistic label for boredom (Table 2).

Finally, for each emotion, Chi square tests revealed no differences due to age.

**Table 2.** Frequency (percentage) of achievement emotions named accurately, Chi square tests (degrees of freedom, $df$), and significance levels ($p$)

| Achievement emotions | Frequency (%) | $X^2$ ($df$) | $p$ |
|---|---|---|---|
| Enjoyment | 52 (98%) | 49.075 (1, 53) | < .001 |
| Pride | 26 (49%) | .019 (1, 53) | .891 |
| Hope | 35 (66%) | 5.453 (1, 53) | .020 |
| Relief | 41 (87%) | 3.189 (1, 53) | .074 |
| Relaxation | 20 (38%) | 20.447 (1, 47) | < .001 |
| Anxiety | 42 (79%) | 18.132 (1, 53) | < .001 |
| Anger | 52 (98%) | 49.075 (1, 53) | < .001 |
| Shame | 39 (74%) | 11.792 (1, 53) | .001 |
| Boredom | 10 (19%) | 20.547 (1, 53) | < .001 |
| Hopelessness | 53 (100%) | - | - |

## 4.3    Discussion

The analyses of the participants' responses in the naming task revealed, with no differences due to age, that some faces were perceived as not corresponding to the hypothesized achievement emotional labels: pride, hope, and relief among the positive emotions and boredom in particular among the negative emotions.

# 5    General Discussion and Conclusions

The main aim of this work was to test a preliminary version of a pictorial instrument to assess achievement emotions, with Pekrun's control-value theory as the theoretical framework [19]. Despite the role played by emotions in learning contexts and their influence on cognition, motivation, and behavior [17, 20], the literature has only recently paid attention to a wide range of emotions linked to achievement situations [19], specifically in children [for exceptions, see 16, 22]. In addition, in studying recognition of affective stimuli, many research works on emotions focused mostly on basic rather than complex emotions, very frequently using photographs [e.g., 6, 11].

To fill such gaps in the literature, we developed a first version of an instrument representing ten achievement emotions [13] through faces, drawn mainly according to criteria described in Ekman and Friesen' studies [8, 9]. Both in the agreement and matching task, the participants' responses were very accurate, indicating that the proposed drawings adequately corresponded to the hypothesized achievement emotions. These results suggest that the faces could be effectively used, together with the corresponding linguistic labels, as an instrument to evaluate achievement emotions. Presentation of drawings (also when associated with emotional labels) could actually be a privileged way to access the emotional information stored in individuals' networks [11], particularly in computer-based environments often characterized by time constraints. However, the analyses of the responses in the naming task – which requires more elaborate abilities than the other two tasks [2] – revealed some difficulties concerning complex emotions such as pride, hope, relief, and boredom in particular. In-depth analysis of error types may help draw modified versions of the pictorial representations of these emotions to be tested in further studies.

On the whole, the absence of dissimilarities between different age children and adults (except for children's higher agreement on pride and boredom in Study 1) confirms the adequacy of the pictorial representations.

From an applicative perspective, the proposed instrument could be utilized as a TEL product in a variety of ways in the assessment of achievement emotions, ranging from conceptualizing them as antecedents, core characteristics, or consequent processes of any learning program. The instrument has been developed for children, but it could also be used with adults and individuals with special needs, after future studies aiming at evaluating its generalizability.

# References

1. Albanese, O., Molina, P.: Lo sviluppo della comprensione delle emozioni e la sua valutazione. La standardizzazione italiana del Test di Comprensione delle Emozioni (TEC). Edizioni Unicopli, Milano (2008)
2. Battistelli, P.: La comprensione delle emozioni. In: Marchetti, A. (ed.) Conoscenza, Affetti, Socialità. Verso Concezioni Integrate Dello Sviluppo, pp. 147–168. Raffaello Cortina, Milano (1997)
3. Cutting, A., Dunn, J.: Theory of mind, emotion understanding, language and family background: Individual differences and interrelations. Child Development 70, 853–865 (1999)
4. De Sonneville, L.M.J., Verschoor, C.A., Njiokiktjien, C., Op het Veld, V., Toorenaar, N., Vranken, M.: Facial identity and facial emotions: Speed, accuracy, and processing strategies in children and adults. Journal of Clinical and Experimental Neuropsychology 24, 200–213 (2002)
5. Denham, S.A., Couchoud, E.A.: Young preschoolers' understanding of emotions. Child Study Journal 20, 171–192 (1990)
6. Egger, H.L., Pine, D.S., Nelson, E., Leibenluft, E., Ernst, M., Towbin, K.E., Angold, A.: The NIMH Child Emotional Faces PictureSet (NIMH-ChEFS): A new set of children's facial emotion stimuli. International Journal of Methods in Psychiatric Research 20, 145–156 (2011)
7. Ekman, P.: Facial expression and emotion. American Psychologist 48, 384–392 (1993)
8. Ekman, P., Friesen, W.V.: The facial action coding system (FACS). Consulting Psychologists Press, Palo Alto (1978)

9. Ekman, P., Friesen, W.V., Hager, J.C.: Facial Action Coding System. A Human Face, Salt Lake City, UT (2002)
10. Fox, E., Russo, R., Bowles, R., Dutton, K.: Do threatening stimuli draw or hold visual attention in subclinical anxiety? Journal of Experimental Psychology 130, 681–700 (2001)
11. Goeleven, E., De Raedt, R., Leyman, L., Verschuere, B.: The Karolinska Directed Emotional Faces: A validation study. Cognition and Emotion 22, 1094–1118 (2008)
12. Gross, T.F.: The perception of four basic emotions in human and nonhuman faces by childrenwith autismand other developmental disabilities. Journal of Abnormal Child Psychology 32, 469–480 (2004)
13. Kleine, M., Goetz, T., Pekrun, R., Hall, N.: The structure of students' emotions experienced during a mathematical achievement test. ZDM. The International Journal on Mathematics Education 37, 221–225 (2005)
14. Larson, R., Lampman-Petraitis, C.: Daily emotional states as reported by children and adolescents. Child Development 60, 1250–1260 (1989)
15. Lazarus, R.S., Folkman, S.: Stress, appraisal, and coping. Springer, New York (1984)
16. Lichtenfeld, S., Pekrun, R., Stupnisky, R.H., Reiss, K., Murayama, K.: Measuring students' emotions in the early years: The Achievement Emotions Questionnaire-Elementary School (AEQ-ES). Learning and Individual Differences 22, 190–201 (2012)
17. Linnenbrink-Garcia, L., Pekrun, R.: Students' emotions and academic engagement: Introduction to the special issue. Contemporary Educational Psychology 36, 1–3 (2011)
18. Lior, R., Nachson, I.: Impairments in judgment of chimeric faces by schizophrenic and affective patients. International Journal of Neuroscience 97, 185–209 (1999)
19. Pekrun, R.: The control-value theory of achievement emotions: Assumptions, corollaries, and implications for educational research and practice. Educational Psychology Review 18, 315–341 (2006)
20. Pekrun, R., Goetz, T., Frenzel, A.C., Barchfeld, P., Perry, R.P.: Measuring emotions in students' learning and performance: The Achievement Emotions Questionnaire (AEQ). Contemporary Educational Psychology 36, 36–48 (2011)
21. Pekrun, R., Stephens, E.J.: Academic emotions. In: Harris, K.R., Graham, S., Urdan, T., et al. (eds.) APA Educational Psychology Handbook. Individual differences and cultural and contextual factors, vol. 2, pp. 3–31. American Psychological Association, Washington, DC (2012)
22. Raccanello, D., Brondino, M., De Bernardi, B.: Achievement emotions in elementary, middle, and high school: How do students feel about specific contexts in terms of settings and subject-domains? Scandinavian Journal of Psychology (in print)
23. Schultz, R.T., Gauthier, I., Klin, A., Fulbright, R.K., Anderson, A.W., Volkmar, F., et al.: Abnormal ventral temporal cortical activity during face discrimination amongindividuals with autism and Asperger syndrome. Archives of General Psychiatry 57, 331–340 (2000)
24. Tanaka, J.W., Wolf, J.M., Klaiman, C., Koenig, K., Cockburn, J., Herlihy, L., et al.: The perception and identification of facial emotions in individuals with autism spectrum disorders using the *Let's Face It!* EmotionSkillsBattery. Journal of Child Psychology and Psychiatry 53, 1259–1267 (2012)
25. Turner, J.E., Schallert, D.L.: Expectancy–value relationships of shamereactions and shame resiliency. Journal of Educational Psychology 93, 320–329 (2001)
26. Weiner, B.: An attribution theory of achievement motivation and emotion. Psychological Review 92, 548–573 (1985)
27. Wishart, J.G., Cebula, K.R., Willis, D.S., Pitcairn, T.K.: Understanding of facial expressions of emotion bychildren with intellectual disabilities of differing aetiology. Journal of Intellectual Disability Research 51, 551–563 (2007)
28. Zeidner, M.: Test anxiety: The state of the art. Plenum, New York (1998)

# Gamify Your Field Studies
# for Learning about Your Learners

Tania di Mascio[1], Rosella Gennari[2], A. Melonio[2], and Pierpaolo Vittorini[3]

[1] DIEI, University of l'Aquila, V.le Gronchi, 30, 67100 l'Aquila, IT
tania.dimascio@univaq.it
[2] CS Faculty, Free University of Bozen-Bolzano, P.za Domenicani, 3, 39100 Bolzano, IT
gennari@inf.unibz.it
[3] MISP, University of L'Aquila, P.le S. Tommasi, 1, 67100 Coppito, L'Aquila, IT
pierpaolo.vittorini@univaq.it

**Abstract.** TERENCE is an FP7 ICT European project that developed a technology enhanced learning system for supporting its learners, who are primary school children, and their educators. In the course of the project, we run field studies with a large number of learners for analysing the context of use of the system. This paper explains why project constraints triggered the gamification of the field studies, as well as how the field studies were conducted. This paper ends by rummaging over the pros and contras of gamifying field studies as in TERENCE.

## 1  Introduction

Constructivism states that learning depends on the specific learners and the context in which they learn. TERENCE is an FP7 ICT European project that is developing a *technology enhanced learning* (TEL) system for supporting primary-school children in learning to read and comprehend texts, in the main context in which this form of learning takes place—schools.

A TEL system that rests on constructivism should be designed so as to be usable and pedagogically effective for its learners, according to the learning context. As there is not a single design methodology that takes care of both the usability and pedagogical effectiveness of a TEL system, the TERENCE consortium mixes two design methodologies: one, the *user centred design* (UCD), that is iterative and places users at the centre of the design process for attaining usability; the other, the *evidence based design* (EBD), that stresses the role of empirical evidence for attaining pedagogical effectiveness. See [5]. The methodologies are used throughout TERENCE: the system is iteratively designed, starting with the analysis of the context with all the TERENCE users, and revising prototypes of the system through evaluations with users again.

More specifically, the TERENCE analysis of the learning context is concerned with (1) the characteristics and preferences of the TERENCE learners, (2) the learning tasks and their organisation into a stimulation plan by domain experts for the TERENCE learners, (3) the environment. Now, the main sources for learning about the characteristics and preferences of the learners are the learners themselves. Thus the TERENCE consortium planned field studies in UK and Italy with c. 500 learners for "learning about learners", specifically, their characteristics and preferences for the system design.

However, the TERENCE learners are children, aged 7–11 year old. There are a number of data gathering methods for interacting with learners that are adults, however, the same methods cannot be used when learners are young children [4]: for example, [19] explains that children might become anxious at the thought of taking a test, and test taking may conjure up thoughts of school. Hanna et al. [11] give suggestions for interacting with children, in particular, they suggest that *you should not ask children if they want to play the game or do a task, that gives them the option to say no. Instead use phrases such as "Now I need you to..." or "Let's do this..." or "It's time to play...". However, the better thing is playing with them.* Druin [7] moves along the same lines, and suggests to use indirect methods, see Table 1, to interact with children when children play the "user" role. More in general, co-design offers a series of suggestions for gathering data with children as users, where co-design is defined as "collective creativity [...] applied across the whole span of a design process" [20].

**Table 1.** A Comparison of the relationship to adults for each children role [7]

| Role of child | Relationship to adults | | | |
|---|---|---|---|---|
| | indirect | feedback | dialogue | elaborate |
| User | X | | | |
| Tester | X | X | | |
| Informant | X | X | X | |
| Design partner | X | X | X | X |

When situated at school and within school activities, however, co-design has some limitations, in particular, if it is done with many learners and strict timings. See [5] for the organisational school constraints of TERENCE. For instance, schools may impose that all children of a class are involved at the same time in the data gathering, as well as that the timing of the gathering is less than a given time. Such constraints place severe limitations on the data gathering methods one can use in a project. In order to overcome such constraints, one can try to engage classes of learners as best as possible in the data gathering, so as to optimise the time constraints and the quality of the gathered data. One way for engaging classes of learners is to gamify the data gathering. Gamification is the usage of game concepts from game design in order to engage learners and solve problems in non-game situation. Therefore, in TERENCE, we planned the data gathering for the context of use (which is not a game situation *per se*) by using gamification. More specifically, we gamified field studies, borrowing and adapting methods from co-design.

This paper starts outlining the essentials of co-design and gamification and, stronger with that, moves on outlining how we gamified the TERENCE field studies for the context of use analysis. The paper ends by briefly assessing the pros and contras of our approach.

## 2   The Essentials of Co-design and Gamification

This section serves to outline the essence of co-design and gamification, necessary for the remainder of the paper.

## 2.1 Co-design Overview

Co-design [20] evolves from cooperative design and participatory design. It attempts to actively involve all users in the design process in order to help ensure that the product under design meets the users' needs, and is usable. Involvement of users early in the research and ideation phases of the design of a new product is often equated to "asking users what they want". However, therein, the key and the main goal of a cooperative session is the collaboration between users for supporting anybody to imagine, express and access their experience and expectations [21]. Co-design sessions can allow us to create a shared understanding and shared language between participants and the designers so as to understand the new product from the point of view of the participants [3]. The outputs are sources of both inspiration and information for designers and participants.

Specifically, in the area of co-designing with children, the work of Alison Druin [6,7] has provided many frameworks and methods that allow us to work with children as partners during a product design. Several co-design methods can be used with children at different stages of the product design and the appropriate methods may vary depending on the purpose of the research [8,10,24].

There are also examples of co-design at school, with users that are school learners. For instance, in [23], the authors explore the applications of co-design methods with 7–9 children. In [9] the authors describe the empirical studies conducted with 36 children at home and in a school environment.

However, to the best of our knowledge, there are no co-design studies with hundreds of school learners and strict timings as required in the TERENCE project.

## 2.2 Gamification Overview

From both theoretical and empirical points of view, nowadays learners are usually more motivated to participate in school-class activities if these are shaped like games, e.g., see [13]. Gamifing a school class activity requires to introduce specific game elements in the activity [17].

From a purely game-theoretic view point, the necessary elements for turning an activity into a game are the actions or *moves* of the players, with their *outcomes*, so that an action of the players makes the game progress from state to state.

However, from a motivation theory perspective, those elements are not sufficient for making a game engaging. Other elements of digital games such as points, levels, and rewards are therein considered, and have been used to engage learners as players in formal learning contexts. The authors of [18] propose a motivational model that explains more general key factors of game engagement, which encompass other studies in the field. They overview research findings investigating the correlations between the appeal of games and the psychological need satisfaction that play can provide. The surveyed results demonstrate that at least three factors make, in the short term, independent contributions to game engagement:

- *autonomy*, that amounts to experiencing a sense of choice and psychological freedom in playing games;
- *competence*, that is, an individual's inherent desire to feel effective in playing;

– *relatedness needs*, satisfied when learners experience a sense of communion with others.

Autonomy, competence and relatedness needs can be realised by means of diverse game elements. Autonomy can be provided by allowing the player to take decisions, for instance, concerning the player's game levels to play, game avatar or game scenario. Competence is generally realised by carefully balancing the game challenges to the players' skills, providing motivating rewards and feedback. Relatedness needs can be satisfied by allowing collaboration, cooperation or competition, for instance, by means of a personal guide in the form of an avatar or by playing with or against other peers.

## 3  The Gamified Field Study

This section contains a description of the gamified field study in the TERENCE project conducted in Italy in the middle of 2011. The main goal of this field study was the definition of the classes of users representing the starting point of the SW engineering process of the TERENCE ALS. Following UCD practices, see e.g., [12] and [16], we conducted experiments using user-based criteria [22]. The TERENCE data gathering was run as part of the regular school activities in UK and Italy from May to July 2011. The studies involved 2 schools in UK and 6 in Italy, for a total of 293 learners in Italy and 226 in UK. Learners, deaf and hearing, were aged 7–11 year; for details, see Table 2.

Like in co-design, the data were gathered class per class, with c.a 20 children per class, two facilitators and the school class teacher, working as informant for the facilitators and familiar referent figure for children. See [14] and [15]. Due to project organisational constraints, the data gathering with each school class could not last longer than 1 hour.

Despite the number of learners in each school and the strict timing, we aimed at gathering high quality data from learners: we needed genuine and dependable information from children concerning their characteristics, environment and life-style for profiling the learners for the TERENCE system (see [1]). In order to gather high quality data, the

Table 2. An overview of the learners participating in the field study

| Country | Age | Hearing Learners | Deaf Learners | Total |
|---------|-----|------------------|---------------|-------|
| UK | 7 | 41 | – | 41 |
| IT | 7 | 55 | – | 55 |
| UK | 8 | 43 | 13 | 56 |
| IT | 8 | 117 | 5 | 122 |
| UK | 9 | 44 | 11 | 55 |
| IT | 9 | 35 | 1 | 36 |
| UK | 10 | 49 | 12 | 61 |
| IT | 10 | 35 | 8 | 43 |
| UK | 11 | – | 13 | 13 |
| IT | 11 | 37 | – | 37 |
| Total | – | 456 | 63 | 519 |

data gathering was gamified so as to engage the learners as best as possible and comply with the time constraints. The protocol of gamified activites was checked and assessed with school teachers so as to meet the needs of the school learners and constraints. For instance, if a challenge was deemed too difficult or too boring for a school class, it was then revised according to the teachers' feedback.

The data gathering was organised as 6 different game challenges, and each of these was organised a self-referential independent game. There were 2 collaborative games, involving all class learners at the same time, and 4 single-player games. A framework was created for each challenge specifying the goal of the challenge, its moves, and how autonomy, competence and relatedness needs are pursued. Tables 3 and 4 are two instantiations of the framework for one collaborative game and one single-player games.

**Table 3.** Game challenge for learning about how learners make their homework

| |
|---|
| **Type:**<br>collaborative game.<br>**Goals:**<br>the goal of the challenge is to obtain information on how learners make their homework.<br>**Moves:**<br>learners received three set of cards; sets contain cards representing (1) mom, dad, teachers, friends etc. as the persons who the learners make their homework wit; (2) kitchen, living room etc. as the places where learners make their homework; (3) 1 hour, 2 hours, 3 hours etc., referring to the time that learners spend on their homework everyday. The Learners were then asked to put the cards with their responses into the containers. Facilitators ask one learners to extract cards and describe the results. Finally, the entire class discuss the results.<br>**Autonomy:**<br>each learner can choose the cards and the participation to the game; each learner can choose what to tell about the made choices<br>**Competence:**<br>each learner can express their verbal skills.<br>**Relatedness needs:**<br>each learner can feel part of the class by telling about his/her choices, or listening to others' choices. | 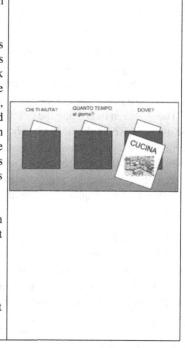 |

At the start of each game challenge, the facilitators explained the goal and the learner's moves for advancing through the game. Autonomy, competence and relatedness needs were pursued across the various challenges. Autonomy was elicited by allowing the learners to choose among several options for tackling a challenge or to take the decision to skip it. Competence was pursued by stimulating diverse skills across game

**Table 4.** Game challenge for learning about how learners interact with their own parents

| |
|---|
| **Type:**<br>single-player game.<br>**Goals:**<br>the goal of the challenge is to obtain information about the interaction between learners and their parents.<br>**Moves:**<br>Learners received two sheets, one with a picture of "mom" in the middle, the other one with a picture of "dad". They were asked to list six (or less or more) activities they often do together with their mom or dad.<br>**Autonomy:**<br>each learner can choose the activities they do with parents and the participation to the game.<br>**Competence:**<br>each learner can express their textual skills.<br>**Relatedness needs:**<br>each learner can feel part of the class by checking his/her choices with his/her friends. | 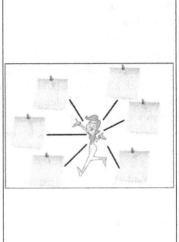 |

challenges, for instance, some games required mainly verbal skills whereas others required mainly drawing skills. The presence of a facilitator working as guidance through the games helped to satisfy relatedness needs; in two challenges, these were achieved by stimulating the school class to work together.

The gathered data using the described assignments have been (1) *stored* in a database for quantitative analysis; (2) *analysed* via statistics $Chi2$ and $Fisher's$ tests (natural variables like gender and age were defined and other dichotomy variables were derived from statistics observations – e.g., rural versus urban); (3) *depicted* via graphics describing the variables associations.

Using tables, graphics and variables associations, we derived four classes here reported:

1. DF (Deaf Female);
2. HF (Hearing Female);
3. H R M N lowAge (Hearing Rural Male North lowAge);
4. H U M N lowAge (Hearing Urban Male Center lowAge);
5. H U M N highAge (Hearing Urban Male Center highAge).

For each class we defined personas used in the TERENCE ALS software development. See [2] for details.

## 4    Conclusions: Pros and Contras of the Method

In the European TERENCE project, we run field studies with young children for analysing the context of use of the TERENCE system. This paper explains why project constraints triggered the gamification of the data gathering activities, as well as how the gamified data gathering was run, borrowing and adapting methods from co-design. In the remainder of this paper, we reflect on the pros and contras of the approach adopted for analysing the context of use.

*Pros.* The data gathered via gamified field studies were qualitatively genuine (a child could express his or her true self), and dependable for creating fine-grained profiles of the learners, also considering the preferences of the TERENCE learners. The reliability of data gathered from learners is supported by evidence gathered from teachers and parents of the involved learners, that was acquired via contextual inquiries. The gamification of data gathering was definitely engaging for children and their teachers to the point that schools became more and more interested in the project, and volunteered to participate in the prosecution of all the TERENCE activities. Moreover, since all children actively participated in the activities under the expert guidance of the facilitator, school constraints for time were respected.

*Contras.* On the other hand, gamification of field studies require considerable human resources and time for constructing material for playing with children. Moreover, the collected data are only semi-structured and therefore their analysis can be long and expensive.

**Acknowledgments.** The authors' work was supported by the TERENCE project. TERENCE is funded by the European Commission through the Seventh Framework Programme for RTD, Strategic Objective ICT-2009.4.2, ICT, Technology-enhanced learning. The contents of the paper reflects only the authors' view and the European Commission is not liable for it.

# References

1. Alrifai, M., de la Prieta, F., Di Mascio, T., Gennari, R., Melonio, A., Vittorini, P.: The Learners' User Classes in the TERENCE Adaptive Learning System. In: Proc. of ICALT 2012. IEEE Press (2012)
2. Alrifai, P.: Deliverable 2.1: Conceptual Model's Specification. Technical Report D2.1, TERENCE project (2011)
3. Ardito, C., Buono, P., Costabile, M.F., Lanzilotti, R., Piccinno, A.: End Users as Co-designers of their Own Tools and Products. J. Vis. Lang. Comput. 23(2), 78–90 (2012)
4. Bruckman, A., Bandlow, A.: The Human-Computer Interaction Handbook: Fundamentals, Evolving Technologies, and Emerging Applications, ch. 1, pp. 1–39. Lawrence Erlbaum and Associates (2002)
5. Cofini, V., Di Giacomo, D., Di Mascio, T., Necozione, S., Vittorini, P.: Evaluation Plan of TERENCE: When the User-Centred Design Meets the Evidence-Based Approach. In: Vittorini, P., Gennari, R., Marenzi, I., de la Prieta, F., Rodríguez, J.M.C. (eds.) International Workshop on Evidence-Based TEL. AISC, vol. 152, pp. 11–18. Springer, Heidelberg (2012)
6. Druin, A.: Cooperative Inquiry: Developing New Technologies for Children with Children. In: Proceedings of the SIGCHI Conference on Human Factors in Computing Systemsonference on Human Factors in Computing Systems, CHI 1999, pp. 592–599. ACM, New York (1999)

7. Druin, A.: The Role of Children in the Design of new Technology. Behaviour and Information Technology 21, 1–25 (2002)
8. Druin, A.: Children as Co-designers of New Technologies: Valuing the Imagination to Transform What Is Possible. New Directions in Youth Development: Theory, Practice, and Research: Youth as Media Creators 128(1), 35–43 (2010)
9. Giaccardi, E., Paredes, P., Díaz, P., Alvarado, D.: Embodied Narratives: a Performative Co-design Technique. In: Proceedings of the Designing Interactive Systems Conference, DIS 2012, pp. 1–10. ACM, New York (2012)
10. Guha, M.L., Druin, A., Chipman, G., Fails, J.A., Simms, S., Farber, A.: Working with Young Children as Technology Design Partners. Communications of the ACM 48(1), 39–42 (2005)
11. Hanna, L., Risden, K., Alexander, K.: Guidelines for usability testing with children. Interactions 4(5), 9–14 (1997)
12. Hartson, H., Andre, T., Williges, R.: Criteria for Evaluating Usability Evaluation Methods. International Journal of Human Computer Interaction 13(4), 373–410 (2001)
13. Jong, M.S., Lee, J., Shang, J.: Educational Use of Computer Games: Where We Are, and What's Next. In: Huang, R., Spector, J.M. (eds.) Reshaping Learning. New Frontiers of Educational Research, pp. 299–320. Springer, Heidelberg (2013)
14. Mascio, T.D.: First User Classification, User Identification, User Needs, and Usability goals, Deliverable D1.2a. Technical report, TERENCE project (2012)
15. Mascio, T.D.: Second User Classification, User Identification, User Needs, and Usability goals, Deliverable D1.2b. Technical report, TERENCE project (2012)
16. Mascio, T.D., Gennari, R.: A Usability Guide to Intelligent Web Tools for the Literacy of Deaf People. In: Integrating Usability Engineering for Designing the Web Experience: Methodologies and Principles, pp. 201–224. ICI Global (2010)
17. Prensky, M.: Digital Game-based Learning. Computational Entertainment 1(1), 21 (2003)
18. Przybylski, A.K., Rigby, C.S., Ryan, R.M.: A Motivational Model of Video Game Engagement. Review of General Psychology 14(2), 154–166 (2010)
19. Rubin, J.: Handbook of Usability Testing. John Wiley and Sons, New York (1994)
20. Sanders, E.B., Stappers, P.J.: Co-creation and the New Landscapes of Design. CoDesign: International Journal of CoCreation in Design and the Arts 4(1), 5–18 (2008)
21. Sleeswijk, V.F., van der Lugt, R., Stappers, P.J.: Participatory Design Needs Participatory Communication. In: Proceedings of the 9th European Conference on Creativity and Innovation, pp. 173–195 (2005)
22. Slegers, K., Gennari, R.: Deliverable 1.1: State of the Art of Methods for the User Analysis and Description of Context of Use. Technical Report D1.1, TERENCE project (2011)
23. Vaajakallio, K., Lee, J., Mattelmäki, T.: "It Has to Be a Group Work!": Co-design With Children. In: Proceedings of the 8th Conference on Interaction Design and Children, IDC 2009, pp. 246–249. ACM, New York (2009)
24. Walsh, G., Druin, A., Guha, M., Foss, E., Golub, E., Hatley, L., Bonsignore, E., Franckel, S.: Layered Elaboration: a New Technique for Co-design with Children. In: Proceedings of the SIGCHI Conference on Human Factors in Computing Systems, CHI 2010, pp. 1237–1240. ACM, New York (2010)

# Learning through Technology

## A Cue-Based Approach

Barbara Giacominelli[1], Margherita Pasini[1], and Rob Hall[2]

[1] Università di Verona, Dipartimento di Filosofia, Pedagogia e Psicologia
Lungadige Porta Vittoria, 17; 37129, Verona
{barbara.gaicominelli,margherita.pasini}@univr.it
[2] Environmetrics Pty Ltd,
PO Box 793 Pymble NSW 2073, Sydney, Australia
rob@environmetrics.com.au

**Abstract.** How can we help people to learn like an expert? The purpose of this study is to describe a method to design work scenarios for a training program using computer software. In particular the focus is on the features that software should have to be useful for learning on the job in the framework of cue-based cognitive theory. There are particular domains, such as aviation, nursing and firefighting, in which decision-making is a fundamental capability. According to Klein [1] who considers decision making a type of skill, in these type of jobs we can try to teach people not only to think like an expert but even how to learn like an expert, providing tools for helping people to achieve expertise in decision making.

**Keywords:** cue-based approach, expertise, ICT, decision making, learning.

## 1 Introduction

In the last 20 years many studies have focused on the cognitive processes used by experts when making decisions, in order to train decision-making skills and help people to achieve expertise. In particular we can classify the literature into three main approaches. The first one, the traditional approach, is aimed at identifying normative models for decision making with the goal of teaching these models in specific training programs. This approach is based on the analytic cognitive processes involved in assessing a situation and making a decision. Driskell, Salas, and Hall [2] analyzed the data from people trained to use analytic decision strategies, the found of the study was that people trained with analytic strategies  resulted in a worse performance than people used unstructured and informal strategies. Means, Salas, Crandall and Jacobs [3] argued that researchers had not successfully demonstrated the efficacy of normative decision training programs. Klein [1] suggested that decision analytic methods are useful only under certain conditions such as when people with low levels of experience work under low time pressure in situations where the relevant factors are well-specified.

T. Di Mascio et al. (eds.), *Methodologies and Intelligent Systems for Technology Enhanced Learning*, Advances in Intelligent Systems and Computing 292,
DOI: 10.1007/978-3-319-07698-0_18, © Springer International Publishing Switzerland 2014

A second approach in the literature of expertise and decision training programs is represented by the heuristics and biases theory of Tversky & Kahneman [4]. Several studies demonstrated the effectiveness of training programs based on this theory [5], [6], but it is not yet clear how these experimental findings can be translated into obtaining the same performance in the real-world environment.

The framework adopted for the present study belongs to a third approach to decision-making processes and associated training programs: the Naturalistic Decision Making approach (NDM) [7], [8]. The NDM approach investigates "the strategies people use in performing complex, ill-structured tasks, under time pressure and uncertainty and in a context of team and organizational constraints" (Klein 1997, p. 340). Klein [1] argues that expert and novice performance can be distinguished by the skills of acquiring, identifying and using cues for diagnosis and response. In particular, cues are presumed to activate a relationship, held in memory, between situational and environmental features and events [9]. Wiggins and O'Hare [10] demonstrated the efficacy of the cues-based approach in diagnosis by using a cue-based training system for weather-related decision-making. In their study participants using a computer-based flight simulator were exposed to a set of cues related to meteorological conditions that could have a bearing on their flight planning. The authors concluded that if the cues that experts use to make a decision are known, it could be possible to construct a decision support system that would help less experienced operators to improve their decision-making capabilities. Subsequently, Wiggins and co-workers explored the influence of cues on the expertise of surgeons and nurses in a hospital context [9]; [11].

The aim of the present study is to extend the cue-based approach in the domain of health care in order to develop a computer training system for improving expertise. This study uses a method to design cue-based scenarios for job simulation and thus inform software developers about the features needed in decision training software in order to capture the cognitive processes used by participants.

## 2    Method

The study aimed to create cue-based scenarios useful for job simulation in a nurse-training program, and involved the nursing staff of neonatal units of two Italian hospitals. We asked to the supervisors of the two units to nominate 4 nurses each, all with very high levels of demonstrated expertise. In total 8 female nurses participated in the study.

### 2.1    Procedure

To create effective scenarios two goals needed to be met. The first required capturing the cue associations relevant to a particular critical situation that the nurses held in memory. The second, required creating scenarios as similar as possible to real-life nursing situations. Furthermore, given that the aim of the study was to create an expertise training program, the scenarios needed to be based on situations requiring quite complex decisions. To reach these goals the Critical Incident Technique (CIT) [12] was used. The CIT is a job analysis procedure aimed to determine which critical factors have made the difference between success and failure in carrying out an important task [12]. In particular, the researcher asked the nurses to describe a situation in their job experience when they were successful and one when they were unsuccessful. Each interview was conducted using a semi-structured interview guide. The nurses were asked to remember the episode and describe it step by step. At each step they were also asked to identify which cues guided their decision and to rate how important each cue was to their decision-making. Both the successful and unsuccessful stories provided useful information about which cues assisted an effective decision and which cues could represent interference in the decision-making process.

Each critical incident interview was used to design a clinical scenario, and at the end of the interviews with the nurses we had 8 scenarios for each hospital unit, in total 16 clinical scenarios.

The last step was to test the scenarios in order to verify if all the nurses agreed on which were the effective decisions for a successful outcome of the scenarios. To verify if the scenarios were generalizable each nurse was asked to analyze the 8 scenarios created by the other neonatal unit. We decided to cross the scenarios between the hospital units to determine whether the decision-making process, in a clinical situation, could be influenced by the particular organizational protocol of the hospital.

In total, there was agreement on the effective decisions and associated cues necessary for resolving the clinical issues in 9 of the scenarios.

## 2.2    Results

Using the 9 scenarios identified in the CIT interviews, and following Loveday et al. [11], three types of experimental task were designed, each one aimed to explore different features of a nurse's decision-making process. In total we developed 3 Feature Identification Tasks (FIT), 3 Feature Discrimination Tasks (FDT), and 3 Transition Tasks (TT).

FITs are developed to train the initial recognition of a critical condition and restrict the search for further cues [13]. In particular, participants are presented with a scenario with relevant cues and are asked to indicate as soon as possible which cue should inform a decision. This process helped to create realistic scenarios to simulate the real job environment. For the present study, a bedside monitor simulator was used to present cues.

FIT example:
Scenario: "Emma, 8 days old, arrives in neonatal intensive care with a suspected bronchioles."
Task: "As quickly as possible indicate the abnormal parameter on the bedside monitor" (Figure 1 shows the monitor for this example).

This type of task is implemented using specific training software able to register and measure the accuracy and reaction time of cue recognition. The participant typically answers using a mouse and clicking on the abnormal parameter.

In contrast, FDTs are designed to train a nurse to discriminate important from irrelevant cues in a critical situation, improving diagnostic capability.

FDT example:
Scenario: "A 3 months old baby arrives in the unit. He presents with agitation and inconsolable crying. The mother says he used to be a very calm and quiet baby. She is mainly worried about the crying. During the triage the mother faints. The baby's cutaneous colour is pink. You can see the bedside monitor with the baby's vital parameters".

In this task, the vital parameters were presented using a bedside monitor to increase the sense of realism.

FDT Task (step 1): "From the options available, determine how you would treat this patient. You may only select one response".

In this task the participant was provided with a list of clinical decision options including the "do nothing" option. The list included the correct option, irrelevant options, and incorrect options.

After the nurse made their choice of action, he/she was asked to indicate the relative importance of the cues they used in making their choice.

FDT Task (step 2): "From the options available, please rate the importance of the different aspects of this scenario in arriving at your response". In this task the participant was provided with a list of all the cues available in the scenario, and he/she was asked to rate each cue on a 10-point scale (1 = "not important at all", 10 = "extremely important").

The FDT can be very useful in a training program if we present it through a software program able to register the decision-making and the rating of the cues.

Finally, the TT task is designed to help develop the skill of using the most effective sequence of cues during a diagnostic process.

TT example:
Scenario: "Timmy is a 20 day old baby, recovering in an intensive neonatal care unit from breath distress and with a bronchiolitis diagnosis; he is pale and has cold feet. You can see his vital parameters on the bedside monitor"

TT Task (step 1): "In addition to the scenario you will be provided with further information surrounding the event. Try to use only the information you consider important for your diagnosis"

We provided the participants with a list of further information, which in a training program, should be organized in an interactive window with categorized cues that can be consulted by clicking on the category.

TT Task (step 2): this is the same as FDT task step 1

TT Task (step 3): this is the same as FDT task step 1

The task is based on the observation that we can distinguish experts from non-experts by the sequence in which they acquire diagnostic information [11].

Her is an example of a monitor vital parameter simulator that could been used in FIT, FDT and TT:

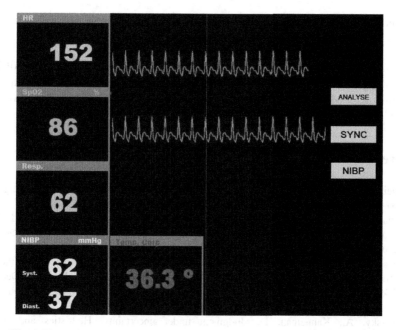

**Fig. 1.** Example of monitor vital parameter simulator usable in FIT, FDT and TT

# 3    Discussion

This study was aimed at testing a cue-based training program for decision making in the nursing domain using job simulation. The program, developed to train nurses to achieve expertise, included three sets of three kinds of tasks. Each kind of task was designed to identify the way nurses used cues as the basis for clinical decision making. The study suggests that a cue-based approach seems very effective in developing a decision-making skills training program. Presenting people with tasks and investigating which type of cue they used to make decisions can train them not only to deal effectively with a particular critical situation, but, overall, to develop metacognitive strategies about the utilization of appropriate cues in general. In this way, as suggested by Klein [1], we can train people to learn like an expert. Our preliminary study, in line with previous research, was based on the premise that directing people's attention to the cognitive processes they use in a situation, may train them to apply this metacognitive approach in other situations, and thus assist in activating an ongoing process of learning on the job.

# 4    Future Directions

This preliminary study is a first step in developing an interactive computer-based training program in Italy capable of presenting scenarios and associated tasks using a monitor. Specific software can give us the possibility of extending the study to a representative sample in a specific domain and verifying the effectiveness of the cue-based training program through longitudinal research.

**Acknowledgements.** The first author wishes to acknowledge the hospitality of Professor Mark Wiggins during her period in The Centre for Elite Performance, Expertise and Training at Macquarie University. She also wishes to thank Dr Thomas Loveday of the Centre for his advice and encouragement.

# References

1. Klein, G.A.: Developing Expertise in Decision Making. Thinking and Reasoning 3(4), 337–352 (1997)
2. Driskell, J.E., Salas, E., Hall, J.K.: The effect of vigilant and hypervigilant decision training on performance. Paper presented at the Annual Meeting of the Society of Industrial and Organizational Psychology, Nashville, TN (1994)
3. Means, B., Salas, E., Crandall, B., Jacobs, O.: Training the decision makers for the real world. In: Klein, G.A., Orasanau, J., Calderwood, R., Zsambok, C.E. (eds.) Decision Making in Action: Models and Methods, pp. 306–326. Ablex Publishing Corporation, Norwood (1993)
4. Tversky, A., Kahneman, D.: Judgment under uncertainty: Heuristics and biases. Science 185, 1124–1131 (1974)
5. Koriat, A., Lichtenstein, S., Fischoff, B.: Training for calibration. Organizational Behavior and Human Performance 26, 149–171 (1980)
6. Yeates, J.F., Estin, P.: Training good judgment. Paper presented at the annual meeting of the Society for Judgment and Decision Making, Chicago, IL (November 1996)
7. Flin, R.: Distributed decision making for offshore oil platform emergencies. In: Proceedings of the Human Factor and Ergonomic Society 40th Annual Meeting, Pittsburg, PA. Human Factors and Ergonomics Society, Inc., Santa Monica (1996)
8. Zsambok, C., Klein, G.: Naturalistic Decision Making. Lawrence Erlbaum Associates Inc., Mahwah (1997)
9. Wiggins, M.: Cue-Based processing and human performance. In: Karwowski, I.W. (ed.) Encyclopedia of Ergonomics and Human Factors, pp. 3262–3267. Taylor and Francis, London (2006)
10. Wiggins, M., O'Hare, D.: Expert and Novice pilot pereptions of static in-fight images of weather. International Journal of Aviation Psychology 13(2), 173–187 (2003)
11. Loveday, T., Wiggins, M., Searle, B.J., Festa, M., Schell, D.: The capability of static and dynamic features to distinguish competent from genuinely expert practitioners in pediatric diagnosis. Human Factors 55(1), 125–137 (2013)
12. Flanagan, J.C.: The critical incident technique. Psychological Bulletin 51(4), 327–359 (1954)
13. Rasmussen, J.: Skills, rules, and knowledge: signals, signs, and symbols, and other distinctions in human performance models. IEEE Transactions on Systems, Man, and Cybernetics 13, 257–266 (1983)

# A Collaborative Distance Learning Portal Integrating 3D Virtual Labs in Biomedicine

Daniele Landro[1], Giovanni De Gasperis[2], and M.D. Guido Macchiarelli[3]

[1] Central Administration, Core Services Management Area, IT Staff, University of L'Aquila
Via Vetoio (Coppito 2), 67100 Coppito (AQ), Italy
[2] Department of Information Engineering, Computer Science and Mathematics,
University of L'Aquila
Via Giovanni Gronchi 18 - Zona industriale di Pile, 67100 L'Aquila, Italy
[3] Department of Life, Health and Environmental Sciences, University of L'Aquila
Via Vetoio (Coppito 2), 67100 Coppito (AQ), Italy
{daniele.landro,giovanni.degasperis,gmacchiarelli}@univaq.it

**Abstract.** In recent years, there has been a growing interest about new techniques for information exchange in medical context. Academic communities need a collaborative portal with a certain degree of interactivity between speaker and audience to provide new learning opportunities. The objective of the research is, therefore, addressed to provide a modular and user-friendly web portal that allows scientific information exchange and takes into account only the centrality of four essential elements: teachers, learners, learning-process and learning-content. This collaborative system contains also 3D learning objects to simulate highly interactive virtual laboratories. The collaborative portal is designed to provide an acceptable teaching tool to integrate face-to-face learning sessions with thematic web-seminars.

**Keywords:** collaborative systems, distance learning, 3D learning objects, virtual laboratories, virtual worlds, usability, web seminars.

## 1 Introduction

The fast growth of the Internet and the wide availability of computing resources raised a great interest towards new communication strategies and new techniques for information exchange in biomedical context [8]. The Web 2.0, is a word widely used to identify a set of technologies increasing the interaction among web users and nowadays it can be efficiently applied in the Academic framework, including biomedical and health education [2] [3] [4].

University education and scientific community use different ways to communicate, such us lectures, seminars, round table, work-shops, labs, etc., which often require a certain degree of interactivity between speaker and audience; this is remarkable in education systems based on Evidence Based Medicine or even Evidence Based Practice [7] [16]. In the research, we faced the problem of providing remediation teaching programs for underperforming medical students [11]. Nowadays, both

T. Di Mascio et al. (eds.), *Methodologies and Intelligent Systems for Technology Enhanced Learning*, Advances in Intelligent Systems and Computing 292,
DOI: 10.1007/978-3-319-07698-0_19, © Springer International Publishing Switzerland 2014

Academics and Students cannot easily benefit distant learning management platforms, mainly due to the need for an adequate training of the operators, the time consuming procedures required to manage multimedia tools as well as the elevated costs [10]. In biomedical and health education, we perceived the need for a user-friendly and low cost tool for distance learning and scientific communications, which may manage several multimedia and hyper-textual tools, as well as virtual labs and podcasts [8].

The practice of medicine has changed dramatically in the past decade because of the increasing demand of a new way to envisage health care, including rapid advances in biomedical knowledge and its application to the practice of medicine. This demand for Continuing Medical Education requires new system for CME, in order to reduce travel costs and optimize time and communication equipment [23].

In this context blended e-learning, obtained integrating face-to-face courses and web interactions, may offer a great door to health education. As a matter of fact, students quite often face on the job training issues, that can be solved only matching epistemological access together with basic information and reasoning skills [15] [6]. A good example of these techniques has been applied in the European Congress of Pathology (ECP); it introduced a preliminary web site to create a Case Database with a live preview [21]. The ECP Case Database, which can be searched and browsed, currently consists of many case presentations from the ECP congresses, years 2006-2012.

## 2     The Collaborative System

Our research was focused on developing a collaborative system to allow academic units within medical schools, health care centers and professional societies to cooperate and provide learning activities, shared materials, web seminars and virtual labs because we strongly believe that a collaborative approach can boost teacher' and doctors' know-how advance and greatly improve medicine practice. In addition, a number of different e-learning objects will play fundamental role in the improvement of CME and academic courses: all members of the learning community are requested to collaborate and share materials using standard IT technologies, without moving from their working place, with no travel and no cost. Our system is easy to use and the technology should be as invisible as possible, just another tool that instructors can use to effectively convey the content and interact with students [1]; in fact each learning framework has four essential elements: teachers, learners, learning process and learning content. The system developed takes into account the centrality of these elements, applying close technological and methodological constraints. Last year, within the European TEMPUS project, we conducted a beta test, involving foreign students, in order to investigate the quality of the portal.

A first version of the web seminars and webinars collaborative platform has been released and will be used in the next academic course of anatomy to carry out the impact assessment.

## 2.1    System Architecture

The web seminar portal is built on client-server architecture in LAMP technology using the new HTML5 language to guarantee a new, simple and intuitive interaction, as shown in Fig. 1.

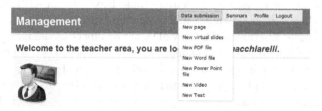

**Fig. 1.** Web Portal Management GUI

Web pages are compliant with the directions of the W3C for accessibility and usability of web portals. During implementation phase, starting from a standard model of the Internet, the project focused on the educational content, creating highly interactive learning processes through a sophisticated and high versatile management system based on learning objects and users' role. Teachers are requested to take care of content preparation split in learning objects, while learners can simply access the platform from any location connected to the Internet.

Now we're developing new features to integrate a laboratory simulator and an adaptive engine to improve system capabilities.

## 2.2    Virtual Labs

One of the more advanced features of the system consists of the possibility to realize three-dimensional highly interactive virtual laboratories, for example to simulate a microscopy laboratory. Virtual worlds are rapidly becoming part of the educational technology framework because they can offer an innovative biomedical education environment to improve learning outcomes [22]. People interact through avatars, personalized by altering shapes, size, skin, hair, and clothing. Avatars can be enlivened with animations to simulate facial expressions, posture, and gestures and can use multiple channels for communication in the virtual world.

Over the past two years, several experiments have been carried out to use protocols and services to simulate three-dimensional virtual worlds, considering the famous world of Second Life as reference [22]. In fact, Second Life is one of the best known of these environments. Now we are developing pilot applications using New World Studio framework [12], an open source multi-platform, multi-language and multi-user 3D application server based on Open Simulator [13]. New World Studio is the server used for the preparation of 3D learning object oriented to laboratory experiments that integrates theory with practical exercises. Also a new browser based on virtual worlds software has become available, exploiting new web standards such as HTML5, CSS3, and WebGL, a JavaScript API to render interactive 3D graphics within any compatible web browser without the use of plug-ins [20]. Cloud Party [5] is one of the first commercial products in this direction.

We are also evaluating to integrate a virtual microscopy simulator inside a learning object, in order to combine practical exercises with theory. A good example of virtual microscope simulator is used in the Virtual Slide Box web site [19]. The web site shows many diagnosis containing a high resolution zoomable image accomplished with a pathogenesis description. The new 3D learning object should integrate a virtual room with virtual humans [9] containing a virtual microscope, which has some buttons to simulate what happens in the real world when people use the instrument.

Our aims were to explore the potential of a virtual world to deliver continuing medical education, to assess the participant learning feedback, to identify the steps required to adapt Virtual Worlds to CME.

## 2.3    Features of the Collaborative System

The web-portal consists of a public area, where users can view some contact information and technical references, send messages and register.

Registered users need to be enabled by the system administrator to join the seminars, according with permissions or payments related. Permission can also be temporally limited. During registration users receive an automatic confirmation email while system administrator is notified about a new registration, to be approved. In case a web user is not interested on registration but only on general information, or on problem reporting, the contact form can be used, to inform technicians by email.

Teachers' area, submitted to previous authorization as well, includes a menu with several features for:

- word, pdf, power point, video files upload;
- simple presentations preparation on the web, with text and images;
- knowledge test preparation;
- 3D learning objects.

3D learning objects are a special and experimental case of a learning objects (such as SCORM packets) that we are developing and testing to define new software components; they represent an interactive three-dimensional digital resource that can be reused and integrated into learning management systems. They contain 3D instructional material, simulation information and metadata; as the time being, there are no standards on this issue, so we are studying a new way to extend SCORM packets to virtual 3D objects.

The system is designed to interact with three actors: learner, instructor, virtual instrument. Each actor has specific roles to interact with the virtual room, according to the operational context.

Fig. 2 shows the collaborative system UML Use Case diagram while in Fig. 3 is designed the sequence diagram used to implement a virtual room, containing virtual instruments, created by the Learning Management System Adaptive Engine (LMSAE). The LMSAE uses innovative technology to assess learners' skills in order to deliver educational materials at the most appropriate and personalized level [18].

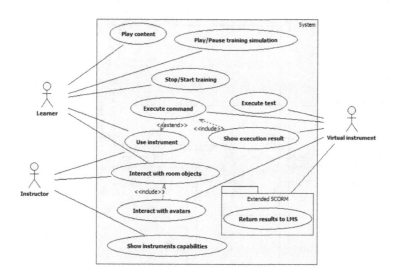

**Fig. 2.** 3D Learning Object system use case diagram (UML)

All simulation results are returned to a localized XML web service which is able to convert and report the results back to a Learning Management System using an extended next generation SCORM package, in this case Tin Can API, that allows learning content and learning systems to speak to each other [17].

The LMSAE will be available in the next academic course of anatomy.

We also are trying to integrate 3D learning objects through a web browser add-on, which extends standard functionality to manage Second Life protocol information exchange and allows users to install specific applications for access and visualization.

The 3D learning object add-on is currently in development state. We confide to integrate, within a year, 3D virtual worlds and virtual microscopy objects in the collaborative web portal.

Each learning object stores author and date. Each teacher can only look at his own documents. Documents archived can be shuffled into seminars, by using wizards, that allows users to select resources and presentation order. Registered users can sign in and view the resources that the administrator has enabled for the visualization.

Videos have a built-in player for playback while Word, PDF, Power Point documents are displayed by the service Google Docs viewer, which only allows content display (not download) through a standard browser.

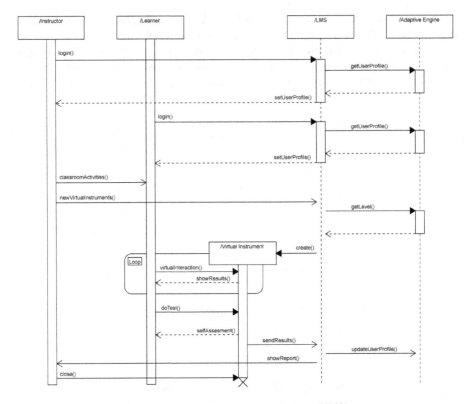

**Fig. 3.** 3D Virtual Lab sequence diagram (UML)

## 2.4    Usability Issues

Portal layout is simple, with bright backgrounds, and user friendly, even for beginners. Interaction interface has a low degree of complexity in order to be used by a heterogeneous public. Page content management has a command bar, with a context menu of available features and a text editor that allows HTML text formatting. Learning content pages have a more complex structure: below the header, there is a video section and the seminar index. A large part of the page is addressed to slide show or virtual laboratory box, with its own toolbar.

We've recently conducted a test program to assess usability; the survey was conducted with eleven participants and the result was a percentile rank of 84%, so it can be interpreted as a grade of an "A". To extend the web portal usability, a better mobile app could be developed using some commercial tools, like Jibe, a new multiuser 3D simulation and virtual world platform created by Reaction Grid [14]. With Jibe you can create, publish and manage your own fully customized virtual world which can be embedded in any web page or accessed from mobile devices. The Jibe platform is an extensible architecture that uses a middleware abstraction layer to communicate with a user database, multiple backend systems and frontends. Current deployments of Jibe worlds utilize the Unity3D web plugin, including an Android version and iOS support under development.

## 3    Conclusions

The study aimed to identify the best ways to develop an innovative web portal which can integrate 3D learning objects into a user-friendly distant learning platform, taking into account the different needs of portal users.

The proposed collaborative system also offers a way to enhance people learning experience, and deepen levels of learners' engagement and collaboration within digital 3D learning environments, reducing user complexity and costs. In order to promote 3D integration into collaborative portal, we're investigating a new SCORM-3D standard. As future usability evaluation, we also plan to compare commercial and open-source 3D virtual worlds solutions in relation to their context of use. Mobile technologies will be assessed during the next development steps.

We expect to assess this portal in the next courses and masters in biomedicine to improve distance learning and take advantages using the integrated adaptive engine, which follows learner attitudes and offers a good extension of face-to-face learning.

## References

1. Barron, A.: A teacher's guide to distance learning. Florida Department of Education, Florida (1999)
2. Boulos, M., Maramba, I., Wheeler, S.: Wikis, blogs and podcasts: a new generation of Web-based tools for virtual collaborative clinical practice and education. Bio Med Central (2006), http://www.biomedcentral.com/1472-6920/6/41 (accessed July 15, 2013)
3. Boulos, M., Wheeler, S.: The emerging Web 2.0 social software: an enabling suite of sociable technologies in health and health care education. Health Information and Libraries Journal (2007), PubMed, http://www.ncbi.nlm.nih.gov/pubmed/17331140 (accessed July 15, 2013)
4. Bryan, A.: Web 2.0: A new wave of innovation for teaching and learning? Educause Review 41, 33–44 (2006), http://www.educause.edu/ir/library/pdf/ERM0621.pdf (accessed July 15, 2013)
5. Cloud Party, http://www.cloudparty.com (accessed July 30, 2013)
6. Funke, K., Bonrath, E., Mardin, W.A., Becker, J.C., Haier, J., Senninger, N., Vowinkel, T., Hoelzen, J.P., Mees, S.T.: Blended learning in surgery using the Inmedea Simulator. Langenbeck's Archives of Surgery (2013), http://www.ncbi.nlm.nih.gov/pubmed/22878596 (accessed July 15, 2013)
7. Ganneboina, J., Revoori, S., Mateti, U.: Implementing awareness program on evidence based medicine for pharmacy students. Journal of Pharmacy and Bioallied Sciences (2007), http://www.ncbi.nlm.nih.gov/pmc/articles/PMC3697198/ (accessed July 15, 2013)
8. Jham, B., Duraes, G., Strassler, H., Sensi, L.: Joining the Podcast Revolution. Journal of Dental Education 72, 278–281 (2008)

9. Jiang, Y., Lim, M., Saxena, A.: Learning Object Arrangements in 3D Scenes using Human Context, Department of Computer Science, Cornell University, Ithaca (2012), http://www.cs.cornell.edu/~asaxena/papers/jianglimsaxena_icml12_arrangement.pdf (accessed July 30, 2013)

10. Macchiarelli, G., Nottola, S.A., Motta, P.M.: Progetto L.I.M.A.: Lezioni Interattive Multimediali per l'Anatomia. In: Cannas, M. (ed.) L'anatomia per Immagini, pp. 135–139. Interlinea Edizioni, Novara (1995)

11. Malau-Aduli, B., Page, W., Cooling, N., Turner, R.: Impact of Self-Efficacy Beliefs on Short- and Long-Term Academic Improvements for Underperforming Medical Students. American Journal of Educational Research (2013), http://pubs.sciepub.com/education/1/6/1/#.UeQ0Lm0lFnU (accessed July 15, 2013)

12. New World Studio, http://www.newworldstudio.net (accessed July 15, 2013)

13. Open Simulator, http://opensimulator.org (accessed July 15, 2013)

14. ReactionGrid, http://reactiongrid.com (accessed July 30, 2013)

15. Rowe, M., Frantz, J., Bozalek, V.: The role of blended learning in the clinical education of healthcare students: a systematic review. Medical Teacher (2012), http://www.ncbi.nlm.nih.gov/pubmed/22455712 (accessed July 29, 2013)

16. Sackett, D.L., Rosenberg, W.M., Gray, J.A., Haynes, R.B., Richardson, W.S.: Evidence based medicine: What it is and what it isn't. British Medical Journal 312, 71–72 (1996)

17. Tin Can API, http://en.wikipedia.org/wiki/Tin_Can_API (accessed July 30, 2013)

18. Tobias, S., Fletcher, J.: Training & Retraining: A Handbook for Business, Industry, Government, and the Military. Macmillan research on education handbook series. Macmillan, USA (2000)

19. Virtual Slide Box, http://www.path.uiowa.edu/virtualslidebox/ (accessed July 15, 2013)

20. WebGL, http://en.wikipedia.org/wiki/WebGL (accessed July 30, 2013)

21. Web Microscope, http://demo.webmicroscope.net (accessed July 15, 2013)

22. Wiecha, J.: Learning in a Virtual World: Experience With Using Second Life for Medical Education. Journal of Medical Internet Research (2010), http://www.jmir.org/2010/1/e1/ (accessed July 15, 2013)

23. Zaragoza-Anderson, K.: Online Webinars for Continuing Medical Education: An Effective Method of Live Distance Learning. International Journal of Instructional Technology and Distance Learning (2008), http://www.itdl.org/Journal/Aug_08/article01.htm (accessed July 29, 2013)

# Using Universal Design for Learning Guidelines to Evaluate a Computer Assisted Note Taking Software Solution

Andrea Mangiatordi

University of Milano Bicocca, Milan, Italy
andrea.mangiatordi@unimib.it

**Abstract.** Universal Design for Learning is a framework proposing a set of guidelines for making learning more accessible to all learners, focusing on the concept of *variability*. By supporting information deconding, strategic thinking and engagement, UDL guidelines provide educators with a methodology for the creation and the evaluation of inclusive lesson plans. This paper discusses the application of UDL principles to the activity of note taking, which is one of the main strategies that can be used to support learning, particularly in secondary and post-secondary education. It requires different skills: capturing and decoding information from instructors is essential, as it is the ability to organize notes effectively. Moreover, note taking requires attention: if a lesson is boring, motivation can drop to the lower possible levels and notes can be totally useless. One specific software package called Evernote is analysed, which allows to write and to capture content for notes in different contexts. The hypothesis is that it can be used in learning contexts adopting a UDL approach. The features of Evernote and of related products are hence examined following the principles and guidelines provided by UDL, with the intent of providing insightful information about its possible use and about its limitations. The analysis shows that even if Evernote can not provide support to all UDL checkpoints per se, it has at least one useful feature for the implementation of each guideline.

**Keywords:** Evernote, note taking, Universal Design for Learning, guidelines.

## 1 Introduction

Universal Design for Learning, or UDL [1], is a framework for instructional content creation and evaluation, addressing the variability of learning. It proposes a set of guidelines based on educational neuroscience research evidence. Brain functionalities are divided in three main domains, connected to an equal number of neural networks: the **recognition network** has the function of collecting and decoding input coming from outside the learner, connecting it to background

T. Di Mascio et al. (eds.), *Methodologies and Intelligent Systems for Technology Enhanced Learning*, Advances in Intelligent Systems and Computing 292, DOI: 10.1007/978-3-319-07698-0_20, © Springer International Publishing Switzerland 2014

knowledge; the **strategic network** manages strategic thinking and problem solving, determining how a learner interacts with the learning environment; the **affective network** connects the inputs and the requests to memories and feelings, encouraging or inhibiting participation, goal setting and engagement.

UDL identifies three main principles addressing the specific functions of each network, suggesting to provide: a) multiple means of representation, b) multiple means of action and expression, c) multiple means of engagement.

From these principles descend nine guidelines, each having a set of checkpoints to be verified by educators preparing or evaluating a lesson plan. The guidelines will be listed and discussed in section 3. UDL implementation basically requires educators to take action in order to lower barriers to learning for their students. In informal learning contexts, however, learners need to develop strategies in order to be effective and need stronger motivation than in formal learning. UDL was initially developed with the specific intent of providing support to students with special needs. Its approach, however, is based on general brain functioning and not on specific intervention on particular impairment, and is actually useful for all students [2].

The activity of note taking is central in supporting learning processes. Taking notes requires different skills, and it is possible to classify them using the above described UDL principles. The *recognition network* captures and decodes information coming from the senses: many note taking systems allow to record such information by various means, like handwriting, OCR, voice recording and speech recognition. Users can decide which tools to combine in order to be effective in capturing information, and often mix a variety of techniques [3]. The *strategic network* is involved when it comes to structuring, organizing and selecting information. Digital note taking tools allow both to organize the content of single notes and to manage collections of notes. Features like full-text search and tagging allow to navigate content in an efficient way. Finally, the *affective network* is involved in the sense that emotional feelings can alter the efficiency in note taking: a boring lesson in a formal learning context can be very difficult to follow, as well as uninteresting information in an informal setting is probably unlikely to be taken into account. Moreover, supporting the affective network also means providing scaffolds to break complex goals into simpler objectives.

The diffusion of the Web led to a significant increase in the amount of information that every person is exposed to on a daily basis. This strengthened the need for note taking systems, and not only for formal learning contexts. Tools like Diigo [4], Microsoft OneNote [5] and Evernote [6], as well as many others, are all suitable for those who need to keep and organize information. In the following sections Evernote will be specifically examined and checked against the guidelines proposed by UDL, in order to describe its possible application in learning contexts. It was chosen as a focus for two main reasons: it is available as a native application on many different platforms (Microsoft Windows, Apple OSX, Apple iOS, Android) and has a web interface for ubiquitous access.

There is also an unofficial client [7] that adds support for GNU/Linux systems. Evernote also offers web browser extensions to format, highlight and capture content, called "web clippers". On some platforms it can also be used together with Skitch [8] to do some basic manipulation and annotation of images.

## 2    State of the Art

Literature about UDL includes both research and position papers, evaluating and discussing the guidelines and the underlying theory. The amount of evidence of positive effects of UDL implementation is limited, yet growing [9]. The main positive results observed involve improvement in the quality of content created by teachers for their classes [10] [11], higher academic results for students with disabilities [12], improvement in consciousness about what barriers exist and how much widespread problems are [13], better feedback from students about the courses they follow [14]. UDL was chosen as a framework for this analysis because, even though it does not strictly require technology in order to be implemented [15], it adds "a philosophical structure to technology-enhanced courses that can change the education landscape and create a more dynamic learning experience for all" [16].

With regard to computer assisted note-taking, a variety of tools is described in literature. Many solutions offer pen-based interfaces, as Classroom Presenter [17]; some, as NoteLook [18] allow the integration of notes and digital video; NotePals [19] provides a way for sharing notes. In recent times, tools designed to quickly store information as notes using cloud storage have multiplied. Evernote has been already mentioned in academic literature in a few works. It appears among a list of useful mobile apps for librarians [20], and is at the center of another study about web-based note capture [21], where its functionality is described and compared to the features of some competitors. Some other papers present research and evaluation directly based on the implementation of Evernote in educational contexts. Students using its mobile features learn in a higher variety of places and tend to annotate more ideas than their peers using only the desktop version [22]. In one case, Evernote was used as a substitute to a paper notebook in a translational science library, and it was found to be a perfect replacement [23].

## 3    Analysis and Discussion

The guidelines proposed by the Universal Design for Learning approach are based on the three main principles described in section 1: educators and instructional designers are encouraged to provide multiple means of representation, action and expression, engagement. In this section the application of these three main ideas is discussed by pointing out possible uses of Evernote (and of the related software, such as Clearly[1], Skitch[2], Peek[3] and Penultimate[4]) that could support such recommendations. The analysis highlights some "merits" and "flaws" for each guideline, to identify how Evernote could be used and what is missing in order to implement the recommendations.

### 3.1    Principle 1 - Provide Multiple Means of Representation

The brain is capable of processing information coming from the senses, even when it is incomplete or distorted [1]. Our ability to identify objects, to connect new information to background knowledge and, in general, to decode input depends on the recognition network.

**Guideline 1 - Provide options for perception.** Learners differ in the way they process information, and can be more sensible to specific formats because of sensorial impairment or simple personal preference. This first guideline suggests to use multimodal communication and to allow learners to adjust content presentation to their preferences.

**Merits:** Evernote supports this by allowing the creation of notes that can include text, images, audio and video. Through Clearly it is also possible to perform on-the-fly styling. This can improve overall readability and provide better reading experience to people with low vision. Images in notes are processed by OCR software in order to make them searchable.

**Flaws:** It is not clear to what extent Evernote clients for both desktop and mobile systems support assistive technology, and specifically screen readers; it is not possible to add alternative text to images from the editor in both web and native apps.

**Guideline 2 - Provide options for language, mathematical expressions, and symbols.** Specialized language can be a barrier to understanding content: this guideline suggests to clarify the use of special symbols, to offer support for vocabulary and to facilitate the comprehension of mathematical notation and formulas. It also recommends to address cross-language issues and to facilitate information structuring.

**Merits:** Evernote can handle MathML formulas if they are already present in saved web pages, so Maths expressions are at least rendered. Native apps can also use system dictionaries to provide vocabulary support to users. The

---

[1] Clearly is a browser plugin, available for Google Chrome, Mozilla Firefox and Opera. It allows to reformat the content of a page on the fly, suppressing marginal content and focusing on the main text only.

[2] Skitch is a simple image manipulation program, providing tools for the annotation of images.

[3] Peek is an iPad application allowing the use of Evernote notebooks as a support for memorization.

[4] Penultimate is an iPad application focused on hand drawing and writing with a stylus.

possibility to structure information by grouping it under tags or notebooks is also present, and notes from saved web pages can be rearranged in order to be better understandable.

**Flaws:** There is no possibility to write expressions from scratch, even though existing ones can be modified.

**Guideline 3 - Provide options for comprehension.** This guideline recommends to activate or supply background knowledge by highlighting patterns, critical features, big ideas, and relationships. It also suggests to guide information processing, visualization, and manipulation, to maximize transfer and generalization.

**Merits:** when a new note is saved from a web page, possibly related notes are listed, to facilitate connections with background knowledge. The same applies to Clearly, that shows related notes at the bottom of a reformatted page.

**Flaws:** there is no real support to transfer and generalization, and pattern highlighting is left to the user.

## 3.2 Principle 2 - Providing Multiple Means of Action and Expression

While the recognition network determines how learners receive information from the environment, the strategic network takes care of how information is elaborated and outputted. The second UDL principle suggests then to diversify the ways in which learners can elaborate on information and produce original content, such as essays, presentations or other forms of expression.

**Guideline 4 - Provide options for physical action.** According to this guideline it is necessary to vary the methods for both response and navigation, while optimizing access to tools and assistive technologies.

**Merits:** Evernote is a cross-platform tool, and this allows users to choose the best solution for their strategic needs. The organization of notes in the UI offers multiple modalities for visual management of collections of notes.

**Flaws:** the overall level of compatibility with assistive technology is uncertain, as already highlighted while discussing Guideline 1.

**Guideline 5 - Provide options for expression and communication.** The use of multiple media is important not only for input, but also for output. To this adds the need for multiple tools for construction and composition, allowing learners to express themselves in the way that mostly suits them. The guideline also suggests the use of graduated levels of support for practice and performance, with the intent of optimizing effort in relation to experience.

**Merits:** using Evernote it is possible to mix information and to organize it, producing new text with the integrated rich text editor. A sharing feature is present in all apps that allows to publish content on various social network sites and also as a public note hosted by Evernote itself.

**Flaws:** There is no feature providing graduated levels of support or scaffolds. The possibility to draw and to write using a stylus is available via Penultimate, but it is limited to iPads.

**Guideline 6 - Provide options for executive functions.** In learning it is important to develop appropriate goal setting, and to support planning and strategy development. It is also necessary to facilitate managing information and resources, while monitoring progress.

**Merits:** a "reminder" feature is present, which can support goal setting, together with the possibility to create checklists in notes. These features are also useful in planning and time management. Notes can be organized in notebooks, categorized using tags, and searched in their full text.

**Flaws:** progress monitoring is only available on one specific platform (iPad), through Evernote Peek.

### 3.3    Principle 3 - Providing Multiple Means of Engagement

The affective network is involved in motivation and engagement in learning: it determines the variability that exists in how individuals can sustain interest in learning or are challenged, and excited [1]. The related UDL principle suggests to offer different means of engagement, to lower emotional barriers and to facilitate the management of personal skills.

**Guideline 7 - Provide options for recruiting interest.** This guideline suggests to optimize individual choice and autonomy, by ensuring discretion in personal customizations. Relevance, value, and authenticity must also be highlighted by contextualizing learning in "real life" situations. One further recommendation is to minimize threats and distractions, creating a safe space for learning.

**Merits:** Evernote is designed for contextual note taking, as it is available for multiple devices and its functions are optimized for fast, unobtrusive collection of information. Also, the Clearly extension allows the reduction of distractions at least while reading web pages.

**Flaws:** the use of Evernote on multitasking devices could lead to continuous distractions.

**Guideline 8 - Provide options for sustaining effort and persistence.** Learning normally implies multiple goals and objectives, and learners may require support in splitting long-term goals in short-term objectives. This guideline also suggests to keep learning challenging by varying demands and resources, while fostering collaboration and community. It is also recommended to provide feedback that guides learners toward mastery, preferring the value of dynamic concepts as effort and practice over static ones as intelligence and ability.

**Merits:** a collaboration feature is present in Evernote that allows to share single notes or entire notebooks, either with selected users or publicly. The above mentioned reminder and checklist features could help in breaking down goals in specific objectives.

**Flaws:** feedback on learning is provided by Evernote Peek, an application only available for iPads.

**Guideline 9 - Provide options for self-regulation.** The last guideline focuses on self-regulation, to be obtained through the promotion of expectations and beliefs that optimize motivation. Coping with anxiety and frustration for ineffective learning can be supported through scaffolds, developing self-assessment and reflection strategies.

**Merits:** also in this case, reminders and checklists can act as scaffold for self-regulation.

**Flaws:** as in the previous guideline discussion, self assessment is only provided as a feature in the Evernote Peek application, which is not cross-platform.

# 4   Conclusions and Future Work

The analysis discussed in the above paragraphs shows that even if Evernote can not provide support to all UDL checkpoints, it has at least one feature that is useful for the implementation of each guideline. The combination of Evernote with tools like Skitch and Clearly extends the compliance of the software package with UDL guidelines. The first two principles of UDL (provide multiple means of representation, provide multiple means for action and expression) are those receiving better support by this note taking system. Besides the emotional effect that the introduction of technology can have on formal learning environments, which is difficult to evaluate *a priori*, some of the features discussed here can support engagement in learning as it is recommended by the UDL framework. If incorporated into a carefully designed curriculum, this particular tool seems then capable of offering a varied and inclusive learning experience. The next step in this research will be testing Evernote and the related tools in a real class environment in which the UDL framework is applied. The test will involve students who are participating in a one-to-one tablet project starting in September 2013. The evaluation will take into account the use of Evernote's features, the quality of the produced notes, the level of collaboration with peers and the impact on academic results measured during the first half of school year 2013-2014 and compared to a control group of the same age.

# References

1. Rose, D., Meyer, A.: Teaching Every Student in the Digital Age: Universal Design for Learning. ASCD, Alexandria (2002)
2. Roberts, K.D., Park, H.J., Brown, S., Cook, B.: Universal Design for instruction in Postsecondary Education: A Systematic review of Empirically Based Articles. Journal of Postsecondary Education and Disability 24(1), 5–15 (2011)

3. Wald, M., Draffan, E., Seale, J.: Disabled Learners' Experiences of E-learning. Journal of Educational Multimedia and Hypermedia 18(3), 341–361 (2009)
4. Diigo, http://diigo.com
5. Microsoft OneNote, http://office.microsoft.com/en-us/onenote
6. Evernote, http://evernote.com
7. NeverNote, http://nevernote.sourceforge.net
8. Skitch, http://evernote.com/skitch
9. Mangiatordi, A., Serenelli, F.: Universal design for learning: A meta-analytic review of 80 abstracts from peer reviewed journals. Research on Education and Media 5(1), 109–118 (2013)
10. Frey, T.J., Andres, D.K., McKeeman, L.A., Lane, J.J.: Collaboration by Design: Integrating Core Pedagogical Content and Special Education Methods Courses in a Preservice Secondary Education Program. The Teacher Educator 47(1), 45–66 (2012)
11. Yang, C., Tzuo, P., Komara, C.: Using WebQuest as a Universal Design for Learning tool to enhance teaching and learning in teacher preparation programs. Journal of College Teaching & Learning 8(3), 21–29 (2011)
12. Coyne, P., Pisha, B., Dalton, B., Zeph, L.A., Smith, N.C.: Literacy by Design: A Universal Design for Learning Approach for Students With Significant Intellectual Disabilities. Remedial and Special Education 33(3), 162–172 (2010)
13. Dymond, S., Renzaglia, A., Chun, E.: Inclusive high school service learning programs: Methods for and barriers to including students with disabilities. Education and Training in Developmental Disabilities 43(1), 20–36 (2008)
14. Katz, J.: The Three Block Model of Universal Design for Learning (UDL): Engaging students in inclusive education. Canadian Journal of Education 36(1), 153–194 (2013)
15. Rose, D.H., Gravel, J.W., Domings, Y.M.: UDL Unplugged: The Role of Technology in UDL. Technical report, The National Center on UDL at CAST, Wakefield, MA (2009)
16. Morra, T., Reynolds, J.: Universal Design for Learning: Application for Technology Enhanced Learning. Inquiry 15(1), 43–51 (2010)
17. Anderson, R., Davis, K.M., Linnell, N., Prince, C., Razmov, V.: Supporting active learning and example based instruction with classroom technology. In: Proceedings of the 38th SIGCSE Technical Symposium on Computer Science Education, pp. 69–73 (2007)
18. Chiu, P., Kapuskar, A., Reitmeier, S., Wilcox, L.: NoteLook: taking notes in meetings with digital video and ink. In: Proceedings of the Seventh ACM International Conference on Multimedia (Part 1), pp. 149–158 (1999)
19. Davis, R.C., Landay, J.A., Chen, V., Huang, J., Lee, R.B., Li, F.C.: NotePals: Lightweight note sharing by the group, for the group. In: Proceedings of the SIGCHI Conference on Human Factors in Computing Systems: The CHI is the Limit, pp. 338–345 (1999)
20. Power, J.L.: Mobile Apps for Librarians. Journal of Access Services 10(2), 138–143 (2013)
21. Ovadia, S.: A Brief Introduction to Web-Based Note Capture. Behavioral & Social Sciences Librarian 31(2), 128–132 (2012)
22. Schepman, A., Rodway, P., Beattie, C., Lambert, J.: An observational study of undergraduate students' adoption of (mobile) note-taking software. Computers in Human Behavior 28(2), 308–317 (2012)
23. Walsh, E., Cho, I.: Using Evernote as an electronic lab notebook in a translational science laboratory. Journal of Laboratory Automation 18(3), 229–234 (2013)

# The Text Simplification in TERENCE

Barbara Arfé[1], Jane Oakhill[2], and Emanuele Pianta[3]

[1] Department of Developmental Psychology and Socialization,
University of Padova, via Venezia 8, 35131 Padova, Italy
[2] School of Psychology, University of Sussex,
Pevensey 1, Brighton, BN1 9QH, United Kingdom
[3] FBK, Via Sommarive 18, I-38123 POVO (TN), Italy
barbara.arfe@unipd.it, j.oakhill@sussex.ac.uk

**Abstract.** In this paper we present the TERENCE system of text simplification. The TERENCE simplification system is intended for use by researchers, educators and policy makers. The method is innovative in the field for two reasons. Firstly, differently from other methods of automatic or manual simplification, it offers a graded, cumulative, simplification of texts. Secondly, differently from other existing methods, it offers a system tailored for two groups of poor comprehenders (deaf and hearing). This paper illustrates the process of text simplification used in TERENCE and presents preliminary results of its testing with elementary school children.

**Keywords:** Readability, text simplification, reading difficulties.

## 1 Text Simplification

Teachers want readers to be able to learn proficiently from texts and they want to supply students with material which is at an appropriate level of difficulty [11,17]. The purpose of text simplification is to provide a reader with a text that is more comprehensible and accessible than the original version [9].

Current systems of text simplification offer a single level of text simplification. Simplification is normally achieved by replacing less frequent words with more frequent and less sophisticated words and by rewriting sentences to make them syntactically simpler. Constructions that are typically simplified in texts for children are passive voice, relative clauses and if-clauses [7,10]. In addition, when texts are for language-impaired readers, conjunctions and anaphors might also be removed [10,18]. This simplification, which is normally automatic and systematic, may result in longer and more fragmented texts, where longer sentences are replaced by multiple shorter sentences. Past research has shown that these texts are not necessarily easier to read [20].

The understandability of a text is more related to its coherence (i.e. its perceived unity) and to the relationship between the elements of the text, rather than simply to the sum of linguistic features of individual words or sentences in the text [12,15]. To

T. Di Mascio et al. (eds.), *Methodologies and Intelligent Systems for Technology
Enhanced Learning*, Advances in Intelligent Systems and Computing 292,
DOI: 10.1007/978-3-319-07698-0_21, © Springer International Publishing Switzerland 2014

improve coherence, the text must be simplified at a cognitive level. TERENCE simplification starts by improving text coherence.

Simplifications at a cognitive level imply the revision of texts at the local level, by increasing the connection between sentences, arranging sentences so that old information precedes new information, and making explicit for the reader important implicit information [5]. However, these simplifications are not always effective. McNamara and colleagues [15,16] have shown that more cohesive texts, that is texts in which links between sentences are made more explicit, require the reader to make fewer inferences and are in general easier to read for low-knowledge readers, who may lack the necessary world knowledge to make inferences, while high-knowledge readers benefit more from low-cohesive texts. That is, high-knowledge readers need some gaps and challenges in the text to be more actively engaged in reading and thus process the text at a deeper level, by making inferences which strengthen their representation of the text meaning [4]. In sum, different levels or kinds of text complexity can be necessary to promote and foster language learning and reading in readers differing for their cognitive and linguistic skills and in their knowledge of the topic.

This idea was the foundation of the entire process of simplification in TERENCE. Developing deaf and hearing children's ability to understand stories is the goal of TERENCE, a EU funded Project aimed at designing an Adaptive Learning System (ALS) for poor readers and their educators. Reading comprehension problems among young hearing and deaf readers can be attributed to deficiencies in a wide range of cognitive and linguistic skills. Deaf children typically show poorer vocabulary than same age hearing children, and have significant problems with syntax. Both deaf and hearing poor comprehenders have difficulty with inferring information that is implicit in the text [1,3,6,19]. These difficulties are related to two types of inferences: those that require the children to make connections between adjoining, or close, sentences in the text, and those that require them to establish more global relations between the information in the text, or to build a mental representation of the text as a whole. Global coherence inferences are however believed to be more problematic for hearing poor comprehenders [6,19]. The system of simplification in TERENCE considers this variety of needs.

## 2    The TERENCE System

The TERENCE system of manual text simplification is applied to a set of 32 stories, written in English and in Italian: the TERENCE repository [2]. TERENCE stories are for primary school children, between the ages of 7 and 11. They constitute, together with games, the primary learning material of TERENCE. In TERENCE, each original story is re-written along three dimensions: global coherence, local coherence, and lexicon-grammar, resulting in three cumulative levels of text simplification. In the broader context of the TERENCE adaptive learning system story levels are selected automatically according to the system's diagnosis of the reader's skills. See also D2.1b, at   http://www.terenceproject.eu/web/guest/public-deliverables.

A driving principle in TERENCE is that gains in reading are acquired through literacy experience and the encounter with authentic and challenging texts. Children's exposure to literary conventions and to well-formed stories represents an important source of literacy experience and of motivation to read. Thus, TERENCE stories are like the texts children find in their textbooks and storybooks. Hence, TERENCE stories are texts written by professional story writers.

The process of text simplification in TERENCE tries to preserve as much as possible the linguistic and textual structure of the original story. Indeed, even children who struggle with reading need to read texts with sufficiently challenging vocabulary and syntax in order to improve their language and reading skills. In line with this principle, and differently from other existing systems, the system of simplification in TERENCE offers readers graded levels of text difficulty, which progressively approach the difficulty that they encounter in the original text. In reading TERENCE stories children are guided to recognize and use the cohesive links and semantic relations in the texts. Attention is placed on the global structure and coherence of the text, so that even the easiest version of the text preserves as much as possible the narrative structure and style of the original story.

The system is the product of an interdisciplinary collaboration between linguists and cognitive psychologists. TERENCE simplifications have the following three goals:

- Improve understandability: the goal of TERENCE simplification is to improve text coherence, taking into account the best possible match between the reader's needs and the characteristics of the text;
- Offer graded texts: different levels and kinds of text complexity can be necessary to promote language learning and reading in readers differing in their cognitive and linguistic skills [15,16]. Types of simplification and text levels are established through an analysis of the readers' cognitive and linguistic needs and on the basis of current models of text comprehension [13]. Text levels in TERENCE are ordered according to their level of difficulty, from 1 (the easiest) to 4 (the most difficult). The number of changes to the texts reflects this graduation, so that texts at Level 3 are the closest to the original, and texts at Level 1 the most distant.
- Minimize changes: authentic language offers children the opportunity to learn in a rich and natural context [14] and simplifications can be detrimental to the natural appeal of children's literature, resulting in less interesting texts, and even making the text more complex and less coherent [20]. Simplifications are thus applied in TERENCE only when they are strictly necessary to improve coherence: e.g. to help the reader to understand main events, or to identify who is performing an important action in the story.

In TERENCE, the story re-writing is cumulative, so that stories are organized in descending levels of difficulty, starting from simplifications to improve global coherence and ending with lexical-grammatical simplifications, as follows [2]:

- **Level 4.** Original story. This is the authentic story, not simplified.

- **Level 3.** Global coherence: the global coherence of the original story is increased, making explicit the information necessary to understand the general meaning of the story, the sequence of events, their location or moral. Making a story more coherent requires that inferential links between main ideas or episodes in the text are made more explicit and/or that links between the text and the reader's (assumed) prior knowledge are made clear. Typically, changes to stories involve both sorts of changes.
- **Level 2.** Local coherence: the text, simplified at the global level, is further simplified at the local level. Improving the understandability of a text also requires establishing more clear and explicit relations between its sentences. Decisions about what links to simplify may depend on the importance of the events described in a sentence or on our hypotheses about what readers know and do not know. Links between events that are peripheral in the story or between descriptive sentences may not need to be simplified. Thus, for example, in TERENCE we do not systematically eliminate ellipsis or pronouns, but reduce their use only when necessary to improve the comprehension of central events in the story [2].
- **Level 1.** Lexicon-grammar: the text simplified at global and local level, is also simplified in terms of its lexicon and grammar, by using more common and concrete words, reducing idiomatic or metaphoric expressions and simplifying syntactically complex sentences. These simplifications can be necessary for those children who might have major difficulties even at the word and sentence level (i.e. deaf children).

A user guide with examples for each simplification level is reported at http://www.terenceproject.eu/c/document_library/get_file?p_l_id=16136&folderId=1 2950&name=DLFE-1927.pdf, and stories are available at http://www.terence project.eu/repository/booken/booken.html.

## 2.1    Testing the TERENCE System: Preliminary Results

TERENCE texts and their simplifications were tested in an exploratory study with a sample of 268 elementary school children, aged 7-11 years, who did not have reading comprehension problems. The goal was to explore preliminarily the effects of simplifications at the various levels within a population of children with average reading skills. According to TERENCE principles of simplification TERENCE stories should be appropriate and engaging for these children too.

**Participants.** Two hundred and sixty-eight Italian children took part in this preliminary study, 146 second (n = 58) and third graders (n =88) (age 7-8 years), 122 fourth (n = 70) and fifth graders (n = 52)  (age 9-10 years). The groups were balanced for gender. All children were native speakers of Italian, and no children presented developmental disabilities or behavioural disorders. Their reading comprehension scores as measured by the MT test [8] were appropriate for their age.

**Procedure and Materials.** Four texts of the TERENCE repository, two intended for the younger group (age 7-8 years) (Story A1 and Story B1), two for the older (9-11 years) (Story A2 and Story B2) were used. Original stories varied in length from 518 to 535 words. Texts intended for the younger children were more explicit and provided more information about the characters and the story events, and so, on average, were longer than the texts for older children. Simplifications at Levels 3 and 2 added explicit information to the texts, and thus Level 3 and 2 texts were longer in length than original texts, up to 586 words in the case of the younger group. However, simplifications were constrained so as not to increase text length excessively, especially in the case of Level 1 texts which were, on average, slightly shorter than Level 2 texts (see Table 1).

**Table 1.** Texts characteristics: mean length in words

|          | Level 4 | Level 3 | Level 2 | Level 1 |
|----------|---------|---------|---------|---------|
| 7-8 yrs  | 526     | 551     | 579     | 568     |
| 9-11 yrs | 494     | 548     | 528     | 527     |

Children were seen on three separate occasions, during a three-week period. Standardized reading comprehension tasks (MT tests) were administered first (week 1) to all children to assess their reading comprehension skills. Then, in two subsequent sessions (week 2 and 3), each child read two stories appropriate to their age group. The first story (e.g. Story A1 or A2) was simplified at one level (Level 1, 2, 3 or 4), the second story (e.g. story B1, or B2), which was read one week later, was simplified at a different level, so that no child read the stories at just one level, and each text was read by about 30 children of the corresponding age group. Story order was counterbalanced.

After reading each story, children answered 12 comprehension questions. The texts were available to the children while they attempted the questions, to allow for re-reading. The 12 comprehension questions assessed children's ability to (1) fill in gaps in the text at a global level (4 inferential questions), (2) link sentences or information that was adjacent in the text (4 inferential questions), (3) retrieve information that was explicitly given in the text (4 literal questions). Thus, the questions tapped global coherence, local coherence and literal understanding.

# 3    Results

Correlations were run to verify whether children's comprehension of TERENCE stories correlated with their reading comprehension skills (MT scores). The correlation was moderate, but significant: $r (255) = .34$, $p < .001$, indicating an association between children's comprehension of TERENCE stories and their reading ability as assessed by a standardized test. All children were all average or good comprehenders and thus their MT scores had limited variance. This might explain why the correlation was not higher.

An analysis of variance with story level (1, 2, 3,4) and age-group (younger vs. older) as between subjects factors shows that children's comprehension increased with simplifications (see Fig. 1). A main effect of story level emerged, $F (3, 247) = 3.59$,

$p = .01$, $\eta^2_{partial} = .04$. As we expected, children already performed quite well on the Level 4 text, that is, on the original text. However, their comprehension of the texts benefited from the simplification, although post-hoc tests (i.e. comparisons between pairs of means) revealed a significant difference only between levels 1 and 4, that is, as a result of the cumulative levels of simplification.

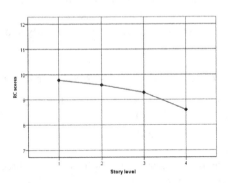

**Fig. 1.** Children's reading comprehension (RC) scores per story level

Older children performed significantly better than younger children, F (1, 247) = 24.11, p < .001, η2partial =.09.

An interaction between story level and group, F (3, 247) = 4.71, p < .005, η2partial =.05, also emerged. As Fig. 2 shows, older children benefited from the simplification, whereas younger children did not.

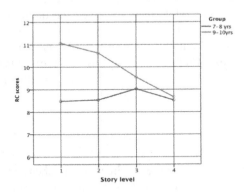

**Fig. 2.** Reading comprehension scores per story level and age group

## 4    Conclusions

In many cases, the reader's difficulty in understanding a text is the result of a lack of sufficient linguistic knowledge. However, more often, readers show a complex range of difficulties. These may include lack of motivation, topic or genre knowledge and lack of cognitive skills, like the ability to monitor their comprehension, use their prior knowledge and make appropriate inferences during reading [1,3,6,18]. In order to be effective, text simplification should address this variety of needs. Besides this, it should offer an authentic opportunity to train reading comprehension processes. This is the idea underlying the entire process of simplification in TERENCE.

Cognitive models and linguistic analysis are integrated in TERENCE to offer a flexible, and intelligent, tool for educational intervention. TERENCE is designed for use in adaptive intelligent technologies and for giving the child the opportunity to experiment reading in tasks of increasing difficulty, modulated according to the expected reader's progress. Preliminary results are encouraging, showing that TERENCE texts are appropriate for 7-11 year old readers, reflect their reading skills and can have positive effects on their reading comprehension. Children enjoyed the texts and were engaged in the reading tasks.

In contrast to previous studies [15,16] we found that even relatively skilled readers benefitted from text simplification. This benefit may depend on the specific characteristics of TERENCE simplification, which preserves as much as possible the style of the natural text [2,14,15], or on the age of the participants. The young readers of this study were still developing their basic reading skills, like inference making and sensitivity to story structure, thus they may have benefited from a text simplification that highlighted important information and logical relations in the text. The results of this study show a slow but regular progression in reading comprehension, mirroring the cumulative nature of the text changes.

The effect of text simplification is not dramatic, though, for these skilled readers, as might be expected [15,16]. Moreover, younger children seem not to benefit so much from the simplification as do older children. Inspection of correct answers at story Level 4 shows that the original story versions were proportionally equally difficult for the two groups (Fig. 2). Thus, the greater sensitivity of older children to the TERENCE simplification cannot be explained by the initial greater difficulty of the texts they read. Another possible explanation is that this lack of effect was related to the length of the texts. Indeed, texts intended for younger children were more explicit and thus longer than those for the older children (Table 1), which contained less explicit information. The further increase in length due to text simplification could have resulted in a cognitive cost for these novice 7-8 year old readers, who were not yet perfectly fluent in reading. That is, the possible gain arising from text simplifications could have been diluted by the demands of reading a longer text. This hypothesis should be tested in future studies, and on poor comprehenders.

**Acknowledgments.** This research was supported by the European Commission through the Seventh Framework Programme for Research and Technological Development. Strategic Objective ICT- 2009.4.2 : ICT : Technology-enhanced learning for TERENCE project.

# References

1. Arfé, B., Boscolo, P.: Causal coherence in deaf and hearing students' written narratives. Discourse Processes 42, 271–300 (2006)
2. Arfé, B., Oakhill, J., Pianta, E., Alrifai, M.: Story simplification user guide. Technical report D.2.2, TERENCE Project (2012)
3. Arfé, B., Perondi, I.: Deaf and hearing students' referential strategies in writing: What referential cohesion tells us about deaf students' literacy development. First Language 28, 355–374 (2008)
4. Benjamin, R.G.: Reconstructing readability: Recent developments and reccommendations in the Analysis of text difficulty. Educational Psychology Review 24, 63–88 (2012)
5. Britton, B., Gülgöz, S., Glynn, S.: Impact of good and poor writing on learners: Research and theory. In: Learning from Textbooks: Theory and Practice, pp. 1–46. Lawrence Erlbaum, Hillsdale (1993)
6. Cain, K., Oakhill, J.V.: Profiles of children with specific comprehension difficulties. British Journal of Educational Psychology 76(4), 683–696 (2006)
7. Carroll, J., Minnen, G., Pearce, D., Canning, Y., Devlin, S., Tait, J.: Simplifying text for language-impaired readers. In: Proceedings of the 9th Conference of the European Chapter of the ACL (EACL 1999), Bergen, Norway (1999)
8. Cornoldi, C., Colpo, G.: Prove di lettura MT per la scuola elementare (Tests of reading MT for primaryschool). Organizzazioni Speciali, Firenze (1998)
9. Crossley, S.A., Allen, D., McNamara, D.S.: Text readability and intuitive simplification: A comparison of readability formulas. Reading in a Foreign Language 23, 86–101 (2011)
10. De Belder, J., Moens, M.F.: Text simplification for children. In: Proceedings of the SIGIR Workshop on Accessible Search Systems, Geneva, July 23, pp. 19–26. ACM (2010)
11. Gunning, T.G.: The role of readability in today's classrooms. Topics in Language Disorders 23(3), 175–189 (2003)
12. Kintsch, W.: The role of knowledge in discourse comprehension: A construction–integration model. Psychological Review 95, 163–182 (1988)
13. Kintsch, W.: The Construction-Integration model of text comprehension and its implications for instruction. In: Ruddell, R., Unrau, N. (eds.) Theroretiocal Models and Processes of Reading, 5th edn. International Reading Association (2004)
14. Larsen-Freeman, D.: Techniques and Principles in Language Teaching. Oxford University Press, Oxford (2002)
15. McNamara, D., Kintsch, W.: Learning from texts: Effects of prior knowledge and text coherence. Discourse Processes 22, 247–288 (1996)
16. McNamara, D., Kintsch, E., Songer, N., Kintsch, W.: Are good texts always better? Interactions of text coherence, background knowledge, and levels of understanding in learning from text. Cognition and Instruction 14, 1–43 (1996)
17. Meyer, B.J.F.: Text coherence and readability. Topics in Language Disorders 23(3), 204–224 (2003)
18. Mich, O., Vettori, C.: E-Stories for Educating Deaf Children in Literacy. Description and Evaluation of the DAMA procedure. Technical Report, LODE project (2011), http://lode.fbk.eu/pubblicazioni.html
19. Oakhill, J.V.: Inferential and memory skills in children's comprehension of stories. British Journal of Educational Psychology 54, 31–39 (1984)
20. Young, D.: Linguistic simplification of SL reading material: Effective instructional practice? The Modern Language Journal 83, 350–366 (1999)

# The Pedagogical Evaluation of TERENCE: Preliminary Results for Hearing Learners in Italy

V. Cofini[1], D. Di Giacomo[1], Tania di Mascio[2], Rosella Gennari[3], and Pierpaolo Vittorini[1]

[1] Dep. of LHES, University of L'Aquila, 67100, L'Aquila, Italy
{vincenza.cofini,dina.digiacomo,pierpaolo.vittorini}@univaq.it
[2] Dep. of EISM, University of L'Aquila, 67100, L'Aquila, Italy
tania.dimascio@univaq.it
[3] Faculty of Computer Science, FUB, 39100 Bolzano, Italy
gennari@inf.unibz.it

**Abstract.** TERENCE is a European FP7 ICT multidisciplinary project that is developing an Adaptive Learning System (ALS) for supporting poor comprehenders and their educators. The paper reports on the evaluation of the pedagogical effectiveness of TERENCE, i.e., its ability of increasing reading comprehension. The evaluation, for hearing children, in Italy, showed the efficiency of TERENCE in the improvement of comprehension ability, by means of silent reading associated with smart games in scholar age.

**Keywords:** TEL, evaluation, reading comprehension, statistical analysis.

## 1 Introduction

Nowadays, about 10% of young children are estimated to be poor comprehenders. They are proficient in word decoding and other low-level cognitive skills, but show problems in deep text comprehension. In particular, studies (see e.g.[3]) experimentally demonstrate that poor comprehenders fail to master the following reasoning skills in processing written stories: (s1) coherent use of cohesive devices such as temporal connectives, (s2) inference-making from different or distant parts of a text, integrating them coherently, (s3) detection of inconsistencies in texts. Experiments show that inference-making questions centred around s1, s2, and s3, together with adequate visual aids, are pedagogically effective in fostering deep comprehension of stories. However, finding stories and educational material appropriate for poor comprehenders is a challenge, and educators are left alone in their daily interaction with them. Most learning material for novice comprehenders is paper based and not easily customizable to the specific requirements of poor comprehenders. Few systems promote reading interventions, but are based on high-school or university textbooks, instead of child oriented stories.

TERENCE [2] is a European FP7 ICT multidisciplinary project that is developing an Adaptive Learning System (ALS) for supporting poor comprehenders and their educators. The project is placed in the area of Technology Enhanced Learning (TEL) and its main objectives are:

– to develop the first ALS, in Italian and in English, for improving the reading comprehension of 7-10 years old poor comprehenders, building upon effective paper-and-pencil reading strategies, and framing them into a playful and stimulating pedagogy-driven environment;
– to evaluate the pedagogical effectiveness of TERENCE, i.e., its ability of increasing reading comprehension.

This paper mainly focuses on the second objective. In particular, the goals of the pedagogical evaluation for hearing learner in Italy, aiming at investigating (i) a pre/post difference in the experimental group (Section 4.3) and in the single schools that made up of the experimental group (Sections 4.1 and 4.2), (ii) a pre/post difference of the experimental group and the single schools with respect to a control group (Section 4.4), and (iii) whether a different effect can be identified in poor-comprehenders and in good-comprehenders (Section 4.4).

## 2   Participants

The evaluation in Italy for the hearing learners counted 526 learner participants, all aged 7–11 year old. The sample is made up of subjects in scholar age dived into two groups: an experimental group and a control group. The experimental group is made up of two schools, i.e., "Avezzano" and "Pescina Centro", while the control group is made up of one school, i.e., "Rieti". Only the experimental group was submitted to the stimulation program, while the control group was enrolled to compare the performances of the

**Table 1.** Italian learner participants of the large-scale usability evaluation

| Group | Location | Class | Number |
|-------|----------|-------|--------|
| Experimental | Pescina Centro | 2A | 19 |
| | | 3A | 18 |
| | | 4A | 22 |
| | | 5A | 15 |
| | | Total | **74** |
| | Avezzano | 2A | 19 |
| | | 3A | 17 |
| | | 4A | 27 |
| | | 5A | 27 |
| | | 2B | 28 |
| | | 3B | 27 |
| | | 4B | 27 |
| | | 3C | 27 |
| | | 4C | 26 |
| | | 5C | 24 |
| | | 4D | 21 |
| | | Total | **270** |
| | | Total | **344** |
| Control | Rieti | 2 | 51 |
| | | 3 | 49 |
| | | 4 | 44 |
| | | 5 | 38 |
| | | Total | **182** |

schools that used TERENCE with a school that didn't. See Table 1 for a recap of the participants, the group they belong to, the school they come from, and their class. The large amount of learners we involved mainly depends on organisational constraints of the project and some constraint introduced by schools; e.g., in order not to introduce discriminations, schools suggested to include in the evaluation all children of a class.

All participants used at least once a PC with mouse in a special classroom and learners were supervised by trained tutors, and we arranged a classroom for each school, bringing in a sufficient number of laptops that allowed all learners to participate simultaneously in the stimulation.

The stimulation lasted 3 months and, in order to guarantee the project time constraint within the allocated human resources, a strict scheduling was designed.

# 3  Tasks and Materials

The "MT prove di lettura" [4] test was applied for the cognitive evaluation. The MT tasks are a measure of the reading ability in scholar age: the test is composed of two trials, an individual one and a group trial. The individual trial evaluates the reading correctness, the group trial measures the comprehension ability. In detail, comprehension ability is assessed by letting the children read a story and answer a set of questions. Depending on the answers, a score ranging from 0 to 10 is assigned. The higher the score, the better the comprehension. Depending on the class, and on the score, a child can be assigned to one of the following clusters: "need for immediate intervention", "attention is needed", "sufficient performance", "complete performance". Accordingly, a poor comprehender is a child belonging to one of the first two clusters. The reading correctness instead regards the production phase, and are thus not investigated in the following, but only reported for completeness.

MT tasks were submitted to the sample by six clinical psychologists unaware about the objectives of the study. All groups underwent to pre- and post-test evaluation. Each subject was evaluated in 40 minutes (both reading and comprehension trials). The cognitive evaluation sessions were conducted in dedicated rooms of educational institutions. One month was needed in order to complete each phase of the evaluation. The pre-test was carried out in January 2013 and the post-test in May/June 2013.

The analyses reported below were performed with the Stata 12 MP software [1].

# 4  Results and Discussion

The section reports the results as follows. Subsection 4.1 describes the "Pescina Centro" experimental group and reports on the pre/post comparison. Subsection 4.2 describes the "Avezzano" experimental group and reports on the pre/post comparison. Subsection 4.3 describes the whole experimental group and reports on the pre/post comparison. Finally, subsection 4.4 describes the "Rieti" control group, reports on the pre/post difference of both the experimental group and the single schools with respect to the control group, as well as whether a different effect can be identified in poor-comprehenders and in good-comprehenders.

It is worth remarking that all learners of the experimental group used TERENCE, though some of them were excluded due to the defined inclusion criteria, i.e., parents must have signed the informed consent, learners must be present during both the pre or post testing, and learners must not have serious cognitive diseases (e.g., children with Down's syndrome).

## 4.1  Pescina Centro

The "Pescina Centro" school is made up 68 students (13.47% of the whole students), divided into 29 females (42.65%) and 39 males (57.35%). Furthermore, 17 students belonged to the second class (25%), 16 students to the third class (23.53%), 20 students to the fourth (29.41%), and 15 students to the fifth class (22.06%).

Table 2 summarises the psycho-pedagogical characteristic investigated by the MT tests. According to the comprehension variable, 14 students (20.59%) resulted poor-comprehenders at the beginning of the experiment, and only 6 (8.82%) at the end of the experiment. Furthermore,

**Table 2.** Psycho-pedagogical variables and pre/post comparisons, by class, in "Pescina centro"

| Variable | Pre/Post | Obs | Mean | Std.Dev. | Min | Max |
|---|---|---|---|---|---|---|
| Comprehension | Pre | 68 | 7.6176 | 1.6930 | 4 | 10 |
| | Post | 68 | 8.2647 | 1.1922 | 5 | 10 |
| Speed | Pre | 68 | 51.691 | 32.412 | 21 | 192 |
| | Post | 68 | 35.764 | 16.526 | 15 | 129 |
| Syllables | Pre | 68 | 2.4573 | 1.0684 | .52 | 4.7 |
| | Post | 68 | 3.1470 | .92846 | 1.28 | 6.4 |
| Reading time | Pre | 68 | 149 | 52.850 | 65 | 272 |
| | Post | 68 | 132.57 | 38.582 | 67 | 240 |
| Correctness | Pre | 68 | 2.7352 | 2.4714 | 0 | 12 |
| | Post | 68 | 1.8529 | 2.0019 | 0 | 10 |

| Class | Comprehension | | z | p |
|---|---|---|---|---|
| | Pre | Post | | |
| 2 | 7.823529 | 8.235294 | -1.932 | 0.0534 |
| 3 | 7.125 | 8 | -2.949 | 0.0032 |
| 4 | 8 | 8.65 | -2.598 | 0.0094 |
| 5 | 7.4 | 8.066667 | -2.226 | 0.0260 |

a Wilcoxon signed-rank test [5] asserted that the difference is statistically significant ($z=-4.904$, $p<0.0001$). The same analysis, divided by class, highlights that the pre/post comparison is not significant – even if borderline – only for the second class.

## 4.2  Avezzano

The "Avezzano" school is made up 254 students (50.30% of the whole students), divided into 131 females (51.57%) and 123 males (48.43%). Furthermore, 44 students belonged to the second class (17.32%), 65 students to the third class (25.59%), 100 students to the fourth (39.37%), and 45 students to the fifth class (17.72%).

Table 3 summarises the psycho-pedagogical characteristic investigated by the MT tests. According to the comprehension variable, 15 students (5.95%) resulted poor-comprehenders at the beginning of the experiment, and only 2 (0.79%) at the end of the experiment. A Wilcoxon signed-rank test [5] asserted that the difference is statistically significant ($z=-2.266$, $p=0.0234$ ). The same analysis, divided by class, highlights that the pre/post comparison is significant only for the fourth class. The lack of significance in this case can be explained by the fact that "Avezzano" starts with an average level of comprehension higher that "Pescina Centro", and close to the maximum. Therefore, a probable ceiling effect took place, i.e., the possibility of a further increase for already skilled learners is minor than learners with an higher margin of improvement.

**Table 3.** Psycho-pedagogical variables and pre/post comparisons, by class, in "Avezzano"

| Variable | Pre/Post | Obs | Mean | Std.Dev. | Min | Max |
|---|---|---|---|---|---|---|
| Comprehension | Pre | 254 | 8.562992 | 1.46162 | 5 | 10 |
| | Post | 254 | 8.767717 | .8511682 | 6 | 10 |
| Speed | Pre | 254 | 37.27953 | 15.6931 | 16 | 116 |
| | Post | 254 | 31.69685 | 11.4204 | 3 | 83 |
| Syllables | Pre | 254 | 2.97748 | .9194232 | .85 | 5.89 |
| | Post | 254 | 3.342953 | .9187147 | 1.2 | 5.7 |
| Reading time | Pre | 254 | 117.4488 | 41.62883 | 58 | 240 |
| | Post | 254 | 124.4724 | 41.11006 | 58 | 240 |
| Correcteness | Pre | 254 | 1.76378 | 2.331963 | 0 | 17 |
| | Post | 254 | 1.755906 | 2.326056 | 0 | 15 |

| Class | Comprehension | | z | p |
|---|---|---|---|---|
| | Pre | Post | | |
| 2 | 8.25 | 8.545455 | -1.598 | 0.11 |
| 3 | 8.784615 | 8.723077 | 0.769 | 0.4421 |
| 4 | 8.51 | 8.91 | -2.962 | 0.0031 |
| 5 | 8.666667 | 8.733333 | -0.120 | 0.9048 |

## 4.3 Experimental Group

The experimental group is the sum of the two aforementioned schools "Pescina Centro" and "Avezzano".

In summary, the experimental group is made up 322 students (63.76% of the whole students), divided into 160 females (49.49%) and 162 males (50.31%). Furthermore, 61 students belonged to the second class (18.94%), 81 students to the third class (25.16%), 120 students to the fourth (37.27%), and 60 students to the fifth class (18.63%).

Table 4 summarises the psycho-pedagogical characteristic investigated by the MT tests. According to the comprehension variable, 29 students (9.06%) resulted poor-comprehenders at the beginning of the experiment, and only 8 (2.50%) at the end of the experiment. A Wilcoxon signed-rank test [5] asserted

**Table 4.** Psycho-pedagogical variables and pre/post comparisons, by class, in the experimental group

| Variable | Pre/Post | Obs | Mean | Std.Dev. | Min | Max |
|---|---|---|---|---|---|---|
| Comprehension | Pre | 322 | 8.363354 | 1.559303 | 4 | 10 |
| | Post | 322 | 8.661491 | .9539269 | 5 | 10 |
| Speed | Pre | 322 | 40.32298 | 21.16795 | 16 | 192 |
| | Post | 322 | 32.5559 | 12.75036 | 3 | 129 |
| Syllables | Pre | 322 | 2.86764 | .9745394 | .52 | 5.89 |
| | Post | 322 | 3.301584 | .9228104 | 1.2 | 6.4 |
| Reading time | Pre | 322 | 124.1118 | 45.99126 | 58 | 272 |
| | Post | 322 | 126.1832 | 40.66565 | 58 | 240 |
| Correctness | Pre | 322 | 1.968944 | 2.391372 | 0 | 17 |
| | Post | 322 | 1.776398 | 2.25887 | 0 | 15 |

| Class | Comprehension | | z | p |
|---|---|---|---|---|
| | Pre | Post | | |
| 2 | 8.131148 | 8.459016 | -2.300 | 0.0214 |
| 3 | 8.45679 | 8.580247 | -0.775 | 0.4383 |
| 4 | 8.425 | 8.866667 | -1.122 | 0.2619 |
| 5 | 8.35 | 8.566667 | -0.120 | 0.9048 |

that the difference is statistically significant (z=-4.200, p<0.0001). The same analysis, divided by class, highlights that the pre/post comparison is significant in the second and fourth class.

## 4.4   Experimental Group and Controls

The control group is made up 183 students (36.24% of the whole students), divided into 88 females (48.09%) and 95 males (51.91%). Furthermore, 51 students (28.02%) belonged to the second class, 49 students to the third class (26.92%), 44 students to the fourth (24.18%), and 38 students to the fifth class (20.88%).

**Table 5.** Psycho-pedagogical variables in the control group

| Variable | Pre/Post | Obs | Mean | Std.Dev. | Min | Max |
|---|---|---|---|---|---|---|
| Comprehension | Pre | 183 | 7.677596 | 1.330079 | 5 | 10 |
| | Post | 183 | 7.546448 | 1.082719 | 5 | 10 |
| Speed | Pre | 183 | 46.69005 | 31.82957 | 18 | 285.71 |
| | Post | 183 | 37.25683 | 14.76574 | 18 | 129 |
| Syllables | Pre | 183 | 2.669051 | 1.111611 | .20625 | 5.39 |
| | Post | 183 | 2.933224 | .7852633 | 1.28 | 5.4 |
| Reading time | Pre | 183 | 136.0328 | 49.41647 | 76 | 240 |
| | Post | 183 | 139.3716 | 36.82202 | 73 | 240 |
| Correctness | Pre | 183 | 1.754098 | 1.806482 | 0 | 10 |
| | Post | 183 | 1.617486 | 1.542839 | 0 | 8 |

Table 5 summarises the psycho-pedagogical characteristic investigated by the MT tests. With respect to the variable under investigation, i.e., comprehension, the control group is more homogeneous to "Pescina Centro" than to "Avezzano", since the average of the comprehension variable, at the pre-test (i.e., 7.67), is closer to "Pescina Centro" (7.61) than to "Avezzano" (8.56). Consequently, the comparisons will be conducted as follows: firstly, the total experimental group will be compared to the control group, with also post-hoc comparisons; secondly, only "Pescina Centro" will be compared to the control group.

The analysis of variance for repeated measures [5] showed that the improvements in the reading comprehension in the experimental groups vs control is statistically significant (F=25.65,p<0.0001). The same analysis, divided by class, shows that the improvements are significant in all classes except the third.

The post-hoc comparison is summarised in Table 6. The correction is done with the Bonferroni method [5]. All comparisons are statistically significant. We now focus on the latest comparison, i.e. Pescina Centro vs Rieti, since these are the two most homo-

**Table 6.** Results of the post-hoc comparison

| Post-hoc | F | p | p-corrected |
|---|---|---|---|
| Avezzano-Pescina Centro | 9.78 | 0.0019 | 0.0057 |
| Avezzano-Rieti | 14.63 | 0.0002 | 0.0006 |
| Pescina Centro-Rieti | 59.44 | 0.0000 | 0.0001 |

geneous groups. As reported in Table 6, the improvements in the reading comprehension in "Pescina Centro" vs control is statistically significant (F=59.44,p<0.0001). The

**Table 7.** Results of the analysis of variance for repeated measures, by class

| Class | Case/control | Comprehension | | F | p |
|---|---|---|---|---|---|
| | | Pre | Post | | |
| 2 | Case | 7.823529 | 8.235294 | 11.41 | 0.0012 |
| | Control | 7.764706 | 7.588235 | | |
| 3 | Case | 7.125 | 8 | 18.61 | 0.0001 |
| | Control | 7.469388 | 7.44898 | | |
| 4 | Case | 8 | 8.65 | 16.86 | 0.0001 |
| | Control | 7.5 | 7.45 | | |
| 5 | Case | 7.4 | 8.066667 | 13.99 | 0.0005 |
| | Control | 8.026316 | 7.684211 | | |

analysis, divided by class (see Table 7), shows that the improvements are significant in all classes.

Finally, Table 8 summarises the average values of the comprehension variable for poor comprehenders and for good comprehenders, as in the pre- and the post-tests. Despite the larger increase in the poor-comprehension group than in the good-comprehension group (0.56 vs 0.06), such a difference is not statistically significant (F=0.82, p=0.3665), mostly because of the small number of poor comprehenders at the post-test (only 8). However, it gives a good indication that the bigger progresses appear in poor comprehenders than in good-comprehenders.

**Table 8.** Descriptive analysis of performances

| Poor comprehender | Pre/Post | Obs | Mean | Std.Dev. | Min | Max |
|---|---|---|---|---|---|---|
| No | Pre | 291 | 8.67354 | 1.256687 | 5 | 10 |
| | Post | 312 | 8.737179 | .8495293 | 7 | 10 |
| Yes | Pre | 29 | 5.310345 | .6037649 | 4 | 6 |
| | Post | 8 | 5.875 | .3535534 | 5 | 6 |

# 5   Conclusions

The paper briefly presented the TERENCE project and delved into the pedagogical effectiveness evaluation that took place in the last months of the project. In summary, the evaluation showed that statistically significant differences exist in the pre/post test, and also with respect to a control group. Our results showed the efficiency of TERENCE in the improvement of comprehension ability by means of silent reading associated with smart games in scholar age. The data evidenced the functional efficiency by integration of smart technology and cognitive development using dedicated educational plans. It is worth remarking that a limit of the experiment is the absence of randomization. However, we could not adopt a randomisation given the firm constraint imposed by the schools, i.e., that all children should use TERENCE. However, the adoption of an homogeneous control (from the ex-post verification we showed that the control group is more homogeneous to Pescina Centro) should guarantee that we could ascribe the discovered difference to TERENCE. A final remark is that children that were unable to finish reading all TERENCE stories asked for logins and password to continue using TERENCE at home during the summer break, thus suggesting a good engagement into the stimulating activities proposed in TERENCE.

**Acknowledgments.** This work was supported by the TERENCE project, funded by the EC through FP7 for RTD, ICT-2009.4.2.

# References

1. Stata 12 MP, http://www.stata.com
2. TERENCE Project, http://www.terenceproject.eu
3. Cain, K., Oakhill, J.V.: Children's Comprehension Problems in Oral and Written Language: A Cognitive Perspective. Guildford Press (2007)
4. Cornoldi, C., Colpo, G.: MT prove di lettura. Giunti Editore (1998)
5. Riffenburgh, R.H.: Statistics in Medicine, 3rd edn. Academic Press (2012)

# Business Games and Simulations:
# Which Factors Play Key Roles in Learning

Andrea Ceschi[1,2], Riccardo Sartori[1,2], Giuseppe Tacconi[1,2], and Dorina Hysenbelli[3]

[1] Department of Philosophy, Education and Psychology, University of Verona,
Lungadige Porta Vittoria 17, 37129 Verona, Italy
{andrea.ceschi,riccardo.sartori,giuseppe.tacconi}@univr.it
[2] CARVET, Center for Action Research in Vocational Education and Training,
Lungadige Porta Vittoria 17, 37129 Verona, Italy
[3] Department of Developmental Psychology and Socialization, University of Padova,
Via Venezia 8, 35131 Padova, Italy
dorina.hysenbelli@unipd.it

**Abstract.** The paper reports the results of an empirical study on the effects and impact of a specific business game, which is also a team competition, treated as an innovative teaching tool in learning. The paper starts by introducing business games and simulations as methods able to improve learning experiences and goes on by dealing with the specific business game simulation used for the aims of our research. Considering the most relevant empirical studies on business games and simulations, the following four factors were extracted in order to test their importance for learning: Decision-Making Experience (DME), Teamwork (T), Simulation Experience Satisfaction (SES), Learning Aims (LA). Each construct has been investigated by using a questionnaire administrated to 48 participants of the Stock Market Learning Simulation divided into 10 teams. Results show the importance of these factors in detecting critics aspectal of learning using a business game simulation.

**Keywords:** Business games and simulations, learning.

## 1 Business Games and Simulations: A Review

Business games are role-playing experiences which address economic and financial issues and aim to develop monetary and financial management skills in participants. Students, trainers and workers taking part in this kind of activities approach managerial issues and take decisions based on market strategies. The main educational aim of business games is to develop decision-making skills and confidence with business strategies [10].

Simulations are evolving open-ended situations of particular social or physical realities in which participants take on bona fide roles with well-defined responsibilities and constraints [2]. They reproduce reality in order to create a better training environment.

T. Di Mascio et al. (eds.), *Methodologies and Intelligent Systems for Technology Enhanced Learning*, Advances in Intelligent Systems and Computing 292,
DOI: 10.1007/978-3-319-07698-0_23, © Springer International Publishing Switzerland 2014

Business games and simulations are direct forms of experiential exercises. "Experimental learning" is a dialectical process where all concepts are continually subject to revision and changes through lived experience are possible. Kolb (1984) states that, in addition to simple cognitive processes, the learning experience has an interactive nature, which constrains different levels of learning [12]. The author underlines that "students immersed in doing" develop abilities that are already inside them. The main aspect of the learning process is based on the "concrete experience", thanks to which the person is totally involved. Accordingly, this allows to do new experiences and to develop the knowledge necessary to give birth to new styles of thinking and doing. This part of learning is considered helpful to illustrate the process of learning, overall in the management of problem solving and decision-making [13].

Business games and simulations provide advantages not found in exercises using discrete, static problems. First, they bridge the gap between the classroom and the real world by providing experience with complex, evolving problems. Second, they can reveal student misconceptions and understandings about the content. Third, and particularly important, they can provide information about students' problem-solving strategies [6].

Considering these characteristics, how widespread are simulations and business games? And in what way do institutions use them to improve learning?

A study shows that, in USA, business games are used by 62% of companies for their training programs. Moreover, 97.5% of AACSB International Business Schools and 66% of teachers in schools use simulations to teach business strategies. Participants can do business activities, take decisions on different areas, and increase abilities [3]. Mintzberg (1987) finds that, during a business game, participants are encouraged to develop and use strategies rather than simply apply knowledge and concepts. If participants are busy in the performance for a period of time which is long enough, they can receive benefits for the evaluation of their own strategy [14]. Education through virtual games allows participants to replicate their behavior in reality. Effectiveness of simulations depends on educational roles and learning in a symbolic form [12].

Compared to classical educational techniques, simulations allow variability in the activities. Participants learn through active learning and through personal styles. Groups of participants playing the simulation and applying their strategies to work can create a group style of working where everyone is involved. In this way, a group becomes a team, where a team is conceived as a group of agents adopting the appropriate joint and individual mental attitudes [17]. Finally, business games and simulations stimulate intellective skills and emotive aspects, since they encourage curiosity, competition in front of challenges and pleasure/displeasure when anyone wins or loses.

## 1.1   Important Factors in Business Games and Simulations

One of the most important aspects of games and simulations is that decisions must be made in team. For example, participants in business games must organize a plan, accomplish it, and carry out an elaborated project. All together. This can be a really demanding and challenging task for people taking part in the game, because they may have just theoretical knowledge and never been involved in such kind of experiences. The polarization between theoretical concepts and practical skills can emerge during the training and be the occasion for participants to reflect on it or even to discuss and argue. In any case, games and simulations may help to fill the gap.

Participants in games and simulations may learn to apply theoretical concepts to reality by means of behaviors elicited in a protected context. "Micro-World" is the expression used by Kofman and Senge (1993) to describe the experience of managers involved in games and simulations, where participants can learn how to manage a business in a controlled environment. For business game's participants, identifying a problem (problem-finding), organizing data (problem-setting) and elaborating a strategy to solve it (problem-solving) can be difficult but also constructive processes [11]. Participants seem to develop their decision-making abilities by considering opponent roles and they consider opponent roles because this encourages their attention to be focused on the moves to do [8]. Given these premises, the so-called "decision-making experience" seems to play an important role in simulations and business games, and that's why we decided to consider it in our study.

Building a strategy based on behavioral styles means that every member feels they belong to a team. A team differs from a group in that it is in direct competition with other teams for resources. In order to be successful, participants must be able to draw data from the context, compete with other teams and plan actions based on self-criticism. This learning can happen easily through the participation in games and simulations. The concept of teamwork is often associated to cohesion. Härtel, Härtel, and Barney (1998) have defined teamwork as an essential characteristic of teams which, over time, have developed a history of shared attitudes and behavioral patterns or norms through experiences or events [7]. Working in team means to try to obtain results based on group climate [1, 15]. A positive and serene group climate is usually associated to better performance. On the contrary, if group climate is tense or people in the team are hostile, the performance is likely to be worse. Teamwork is an important construct that should be considered in testing the potential of a simulation that requires team cohesion, and that's why we decided to consider it in our study.

Satisfaction towards the game experience is another important factor able to influence learning. Games and simulations seem to increase the level of involvement and satisfaction in participants through different activities [5]. In teaching based on simulated realities, participants should feel that activities are not simply didactic tools aiming to test knowledge and abilities, but experiences simulating their professional contexts. It is also important to distinguish participants' satisfaction for a new activity from participants' satisfaction for the success of the team. As already said, simulations can result in different feelings [3]. Nevertheless satisfaction seems to be an important construct that should be considered in testing the potential of games and simulations, and that's why we decided to consider it in our study.

Finally, we can say that business games are virtual simulations that can be innovative educational tools involving participants and letting them both feel satisfied and develop new abilities and knowledge. In order to analyze the learning potentials of business games, it is important to test participants' perception of whether the game has reached the learning aims or not. This is also an important factor to be considered in testing the effectiveness of games and simulations, and that's why we decided to consider it in our study.

## 1.2    The Stock Market Learning Simulation

In order to study the importance for learning of the factors presented above, we have considered the project Stock Market Learning Simulation (SMLS), which is an initiative promoted by different European savings banks and European banking foundations. This simulation constituted an opportunity for people in training to invest a virtual starting capital of 20000 euros on the stock market. Teams had 10 weeks to increase the value of their deposit. The trading was based on share prices and real stock exchanges on the major financial markets.

Each trainee was put in a team with a supervisor and took part in making decisions. The trainees received some information (feedback) about their portfolio in the mid-term and at the end of the competition. In order to involve participants, they had to operate frequently. Otherwise, they were disqualified. It is important to know that they had some advice in order to develop investment, so not only the total value of their account was important, but also their ability to earn from investments suggested by trainers. Focusing on sustainability, the simulation aims to promote prudence and far-sightedness among people, by developing long-term strategies to short-term profit-making without losing sight of economic aims.

## 2    Aim and Method

### 2.1    Participants

Participants were 48 Italian students from Economic studies who took part in the business SMLS (age 18-22, 62% females). They were divided into 10 teams.

### 2.2    Material and Procedure

Since task performance times were equal for every team (the simulation lasted 10 weeks) and all the teams at the end of the competition doubled their initial amount of 20000 euros, a questionnaire was developed and administered to participants in order to test the constructs described in the theoretical part: Decision-Making Experience (DME), Teamwork (T), Simulation Experience Satisfaction (SES), Learning Aims (LA). Each construct is measured by four items developed by considering questions used in studies aiming to analyze prominent factors for learning in business games and simulations [3, 4, 5]. Answers were given by using a five-point rating scale, where 1 expressed the minimum appreciation, while 5 expressed the maximum.

## 3    Results

Table 1 reports the descriptive statistics of the total deposit gained by the 10 teams at the end of the competition, in order to show that, after 10 weeks, all the teams doubled their initial virtual amount of 20000 euros, which means that teams achieved good learning and performance.

**Table 1.** Descriptive statistics of the total deposit gained by teams at the end of the competition

|  | Numbers of teams | Minimum | Maximum | Mean | Standard Deviation |
|---|---|---|---|---|---|
| Total deposit | n = 10 | 40581 | 51526 | 44843 | 3097 |

So, since it was not possible to use performance or time to assess the effectiveness of the business game for learning, for every participant a total score for each construct measured by questionnaire was computed, as well as the total mean scores and the Cronbach Alpha reliability for each construct.

DME's mean score is 4.01 (SD = 1.30). The mean score of T is equal to 3.34 (SD = 1.56). SES presents a mean score of 3.94 (SD = 1.28). LA presents a mean score equal to 3.82 (SD = 1.28).

The reliability of the constructs analyzed are acceptable, varying from 0.62 to 0.82.

## 4    Discussion

The construct presenting the highest mean score is DME (M = 4.01, SD = 1.30), probably because participants had the possibility to make several important and strategic decisions during the business game. This probably allowed them to acquire a personal significant decision-making experience and to develop problem-solving skills. Each team developed its strategy and, consequently, testing DME is important to have a team index of the decision-making experience. It is clear that DME is in part related to T (M = 3.34, SD = 1.56), because participants worked in teams and developed financial tactics together. Theoretical argumentation previously presented also sustains that T facilitates game strategies [10, 16].

A related issue concerns critical factors of team climate and their effects on team performance. One useful framework for the study of teamwork through team climate is constituted by the four factors that West and Farr (1990) identified as being central in determining effective team functioning: vision, participative safety, support for innovation and task orientation [18]. According to West and Farr model, participation in the decision processes of groups increases the likelihood that members produce more outcomes of decision and, therefore, that they offer new ideas. Efficient participation must be supported by safe interpersonal conditions (participative safety), while innovating performance also requires the commitment of groups to achieve the highest possible standard of task performance (task orientation) and to offer articulated and enacted support (support for innovation) for attempts at innovation ideas [9].

Also SES reaches a good score (M = 3.94, SD = 1.28). Participants reported that the business simulation game was a good experience in terms of satisfaction and they justified this by asserting that simulation is a good learning system for such aspects as interactivity and innovation. We can hypothesize that, through simulation, participants strengthen their theoretical concepts and applied business strategies based on the theories they have studied at school. In fact, several empirical studies suggest that business games improve learning [3, 6].

Finally, the mean score obtained by LA (M = 3.82, SD = 1.28) seems to demonstrate that participants were able to recognize the aims of the simulation and appreciate the results of the activities in terms of personal learning.

## 5    Conclusion

Evidence obtained from our study allows us to advance some conclusions about games and simulations as teaching tools.

Games and simulations have the possibility to represent, for participants, positive and stimulating experiences. They also seem to improve team cohesion and group spirit. Every education and training activity could benefit from games and simulations. As in real life, even in business games there could be failures resulting from wrong choices, but just wrong choices and their direct consequences can help people to better understand errors and mistakes and develop solutions.

As a matter of fact, such business games as the Stock Market Learning Simulation have both advantages and disadvantages, but, also considering the results of the questionnaire administered, they can be considered as innovative systems that involve participants in experience strategies and decision-making skills.

## References

1. Anderson, N.R., West, M.A.: Measuring climate for work group innovation: development and validation of the team climate inventory. Journal of Organizational Behavior 19(3), 235–258 (1998)
2. De Freitas, S., Oliver, M.: How can exploratory learning with games and simulations within the curriculum be most effectively evaluated? Computers & Education 46(3), 249–264 (2006)
3. Faria, A.J.: Business simulation games: current usage levels—an update. Simulation & Gaming 29(3), 295–308 (1998)
4. Faria, A.J.: The changing nature of business simulation/gaming research: A brief history. Simulation & Gaming 32(1), 97–110 (2001)
5. Faria, A.J., Wellington, W.J.: A survey of simulation game users, former-users, and never-users. Simulation & Gaming 35(2), 178–207 (2004)
6. Gredler, M.E.: Games and simulations and their relationships to learning. Handbook of Research on Educational Communications and Technology 2, 571–581 (2004)

7. Härtel, C., Härtel, G., Barney, M.: SHAPE: Improving decision-making by aligning organizational characteristics with decision-making requirements and training employees in a metacognitive framework for decision-making and problem-solving. The International Journal of Training Research 4, 79–101 (1998)
8. Hirokawa, R.Y., Poole, M.S.: Communication and group decision making, vol. 77. SAGE Publications, Incorporated (1996)
9. Kivimäki, M., Kuk, G., Elovainio, M., Thomson, L., Kalliomäki-Levanto, T., Heikkilä, A.: The team climate inventory (TCI)—four or five factors? Testing the structure of TCI in samples of low and high complexity jobs. Journal of Occupational and Organizational Psychology 70, 375–389 (1997)
10. Knotts, U.S., Keys, J.B.: Teaching strategic management with a business game. Simulation & Gaming 28(4), 377–394 (1997)
11. Kofman, F., Senge, P.M.: Communities of commitment: The heart of learning organizations. Organizational Dynamics 22(2), 5–23 (1993)
12. Kolb, D.A.: Experiential learning: Experience as the source of learning and development, vol. 1. Prentice-Hall, Englewood Cliffs (1984)
13. Kolb, D.A., Lewis, L.H.: Facilitating experiential learning: Observations and reflections. New Directions for Adult and Continuing Education (30), 99–107 (1986)
14. Mintzberg, H.: Crafting strategy. Harvard Business School Press (1987)
15. Nilniyom, P.: The impacts of group climate on creativity and team performance of auditors in Thailand. International Journal of Business Research Publisher: International Academy of Business and Economics 7(3) (2007)
16. Segev, E.: Strategy, strategy-making, and performance in a business game. Strategic Management Journal 8(6), 565–577 (1987)
17. Tidhar, G.: Team-oriented programming: Social structures. Technical Report 47, Australian Artificial Intelligence Institute, Melbourne, Australia (September 1993)
18. West, M.A., Farr, J.L.: Innovation and creativity at work: Psychological and organizational strategies. John Wiley & Sons, Chichester (1990)

# Cooperate Or Defect?
# How an Agent Based Model Simulation on Helping Behavior Can Be an Educational Tool

Andrea Ceschi[1,2], Dorina Hysenbelli[3], Riccardo Sartori[1,2], and Giuseppe Tacconi[1,2]

[1] Department of Philosophy, Education, Psychology, University of Verona and Psychology
Lungadige Porta Vittoria 17, 37129 Verona, Italy
{andrea.ceschi,riccardo.sartori,giuseppe.tacconi}@univr.it
[2] CARVET, Center for Action Research in Vocational Education and Training,
Lungadige Porta Vittoria 17, 37129 Verona, Italy
[3] Department of Developmental Psychology and Socialization, University of Padova,
Via Venezia 8, 35131 Padova, Italy
dorina.hysenbelli@unipd.it

**Abstract.** Over the last years, Agent Based Models (ABMs) have become an important instrument to simulate social complex phenomena. At the same time, they have shown interesting implications for learning activities. To this purpose, we report a simulation on helping behavior carried out by means of an Agent Based Model (ABM) based on four types of different virtual agents: Warm-Glow Cooperators (WG), who give help because it makes them feel better; Gratitude Cooperators (GC), who give help because they previously received help; Cooperators (C), who give help because of both the reasons mentioned above; Defectors (D) who do not give help at all. We explore the pro-social behavior of each type of agents and the system where they live for a certain amount of time in different situations. This specific ABM shows, in the most effective way, why we should increase the level of helping behavior in the population. Furthermore, assuming that giving and receiving help can be both considered positive activities, WG and GC agent strategies should be those who allow to derive the greatest benefit overall. Taking also in account the pedagogical implications of ABMs, the present simulation can be considered as a good instrument to explain dynamics of helping behavior in a virtual society.

**Keywords:** Agent Based Model, Helping behavior, Educational tool.

## 1    Theoretical Introduction to the Model

Agent-Based Models (ABMs) are a simulation modeling technique that uses virtual agents interacting with other virtual agents within a virtual environment and with certain virtual resources. The aim of this innovative instrument is to simulate and predict possible scenarios based on one or multiple established behaviors. ABMs

T. Di Mascio et al. (eds.), *Methodologies and Intelligent Systems for Technology
Enhanced Learning*, Advances in Intelligent Systems and Computing 292,
DOI: 10.1007/978-3-319-07698-0_24, © Springer International Publishing Switzerland 2014

allow to realize flow simulations, organizational simulations and market simulations in such different fields as Natural Sciences, Engineering and Social Sciences [6, 12].

In Social Sciences, ABMs are used to realize different virtual simulations regarding people interaction. One of the most interesting field is probably related to decision-making area and, in particular, to the possibility of creating agents that better simulate human behaviors. Most social and psychological phenomena occur not as the result of isolated decisions by individuals but rather as the result of repeated interactions between heterogeneous individuals over time, and ABMs are the only one type of models that allow doing this.

Currently, this operation is still very complex. Nevertheless, by implementing several basic rules into the agents, we have tried to reproduce some simple behaviors. Starting from two well-known explanations that interpret pro-social behavior (gratitude and warm-glow), we have implemented the ABM we are going to present in this paper. It should describe how the mutual presence of these two different perspectives in virtual agents could influence life in virtual societies.

## 1.1    ABMs for Teaching and Learning

The use of artificial intelligent agents to study and simulate certain human behaviors can be a powerful way to improve teaching and learning as well. Over the years, artificial intelligence creators have spent many efforts in order to create intelligent systems for educational purposes. As stated by Baylor (2002), modeling human behavior allows to create a metaphor of the "human aspects" in an experimental and controlled computer-based environment [4]. In addition, the architecture of the simulation is based on independent objects able to make a research design more flexible. The power of this instrument has been summarized in Table 1, which, as suggested by Baylor (2002), describes some points in favor of educational ABM purposes for the learner and the researcher (or teacher) [4].

An interesting perspective is given by Malone, Lai, and Grant (1997). They suggest that designers in programing agents should create "glass boxes" instead of "black boxes" [13]. In glass boxes, the essential elements of the way agents reason can be seen and modified by learners. This explicit teaching strategy is opposed to build-in the reasoning of agents in systems invisible to learners. Thus, designing agents in this way can improve the reflective thinking processes [5]. Students learn better when complicated social dynamics are presented as diagrams or, more in general, as structured information (as in Figure 1). Watching dynamic events, interpreting them and adjusting agent's actions can be good exercises to manage information processes. For example, Chen and Yeh (2001) have proposed an ABM of 'school' based on an evolving population driven by single-population. The architecture of the study has taken into consideration traders' search behaviors connected to such psychological factors as peer pressure or such economic factors as the standard of living [8].

**Table 1.** Learning and teaching aspects in favor of ABMs

|  | Learner | Researcher\Teacher |
|---|---|---|
| Agent Interactions | He/she can adjust the interactions according to his/her preferences | Agents can be set as independent objects in the system, lending to more flexibility and interactivity |
| Thinking processes | He/she is encouraged to reflect on thinking processes | Researcher could set agents to operationalize specific human aspects |
| Multiple collaborations | He/she has selection of teachers at his disposal and a willing collaborator if desired for the learning process | Through designing agent-based learning environments with multiple agents, it is also possible to develop multiple mentors |
| Environment settings | The learner can take as much time as needed to learn | The researcher/teacher has more control than in a classroom setting over the learning environment and interactions |

The aim of our study is to create a similar learning experience on the importance of helping behavior in a virtual society. Before showing the model, we first introduce some theories related to helping behavior.

## 1.2    Helping Behavior Theories

Helping behavior has been one of the most important research topics since 1975, when Wilson started his first sociobiology studies. Why do people help others? Which are the factors that influence helping behavior? Which models describe it better?

Trivers (1971) states that helping someone who is not related to us is convenient if the favor is repaid in a second moment [16]. Furthermore, Gouldner (1960) believes that reciprocal altruism is an expression of genetic evolution given that, in every culture, at least one norm of reciprocity exists [10]. Two norms of reciprocity have been identified: direct reciprocity, which describes the kind of cooperation emerging in repeated encounters between the same two individuals [2, 14, 16]; indirect reciprocity, which does not have the same constrain; we help someone, who can decide to help another third person, or we are helped by someone and we decide to help another third person [7, 11, 14]. In the present work, we will consider only indirect reciprocity, and the virtual agents who will adopt a pro-social behavior based on this feature will be called Gratitude Cooperators (G).

Another theory assumes that we help people because this makes us feel better about ourselves [1]. As givers, we feel better people, our self-esteem may benefit from the altruistic act and even our reputation might increase. According to this perspective, the benefit of the receiver (which is supposed to be the main goal of an altruistic behavior) becomes of a secondary importance. In our model, this type of givers will be called Warm-Glow Cooperators (WG). Agents who help others for both of the reasons mentioned above will be called simply Cooperators (C). Finally, agents who do not help others at all will be called Defectors (D). Therefore, this model provides for 4 types of agents (G, WG, C, and D) based on two parameters: 1) the probability of initiating and 2) the probability of passing on a pro-social behavior (Table 2).

**Table 2.** The four types of agents in the present model

|  | Initiate | Pass on |
| --- | --- | --- |
| Gratitude Cooperator | No | Yes |
| Warm Glow Cooperator | Yes | No |
| Cooperator | Yes | Yes |
| Defector | No | No |

Obviously, there are other theories related to the reasons why people give help, like the empathic altruism theory [3], the negative state relief model [9] and the empathic joy theory [15], but, for the model we are going to present, we consider reciprocity and warm-glow as the two most important variables to keep the model simple but functioning.

## 2    The Instrument and the Helping Behavior Model

Each of the four type of agents has four main variables:

- Variable Receive ($r$): increases every time the agent receives help.
- Variable Give ($g$): increases every time the agent gives help.
- Variable Help ($h$): decreases every time the agent gives help. Agents move only if $h > 0$. Agents WG and C are the initiators. They have $h > 0$. Not initiators (D and G) have $h = 0$. Agents G and C increase $h$ levels if they receive help by others agents.
- Variable Wasted Help ($wh$): increases in D and in WG when they receive $h$. This because D and WG cannot use the help received by others and, as a result, the $wh$ cannot be used to move the agent.

We decided to give an arbitrary $h$ value to C and WG agents equal to 100 in order to make the system start moving. This value is not unlimited because we try to represent the real world resources which are usually limited. Below (Figure 1) we present the reasoning scheme of a C agent as an example of the functioning system for a virtual agent.

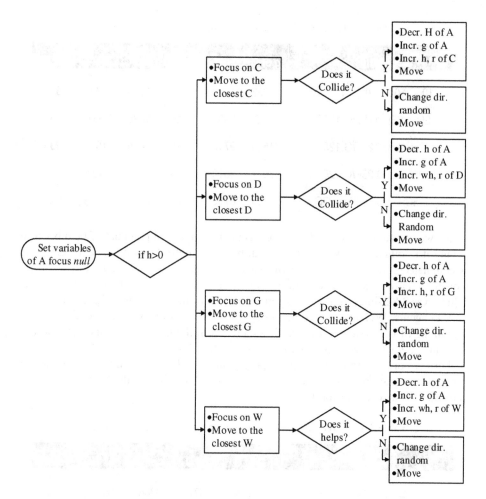

**Fig. 1.** The reasoning system of a C agent

# 3    Results

When the ratios of the four agents are equal (100C:100D:1WG:1G), C agents gave 2.15 times more help comparing to G agents and 1.75 times more comparing to WG agents (Table 3). This because C agents had an initial helping potential ($h = 100$) and they accumulated and reused $h$ when they received it from others, instead G agents had to wait for others' help before being able to give it to others, and WG agents did not accumulate and reuse the help they received from other agents. Additionally, D agents did not give help at all.

**Table 3.** Give and Help amounts for the C, WG, and G agents

| Ratio | g of C | g of G | g of WG | h of C | h of G | h of WG |
|---|---|---|---|---|---|---|
| 100C:100D:100WG:100G | 16632 | 7733 | 9451 | 1854 | 655 | 549 |
| 25C:125D:125WG:125G | 7772 | 6052 | 11835 | 1025 | 531 | 665 |
| 125C:25D:125WG:125G | 21649 | 9718 | 11502 | 1475 | 1056 | 998 |
| 125C:125D:25WG:125G | 17017 | 5913 | 2224 | 1718 | 403 | 276 |
| 125C:125D:125WG:25G | 19169 | 8795 | 11324 | 3144 | 1188 | 1176 |

As we mentioned above, C and WG agents had an initial potential of help because giving it to others makes them feel good, but, differently from WG agents, C agents accumulated further $h$ while receiving help from others and were able to pass it on. For these reasons, at the end of the system's life C agents showed the highest $h$ level which was 2.82 times higher comparing to G agents and 3.37 times higher comparing to WG agents. WG agents had the lowest $h$ level because they started to use it in the beginning of the system's life, but differently from the C and G agents, they did not accumulate $h$ while interacting with others in order to reuse it.

Furthermore, all four agents were programmed to have the same probabilities to receive help from others (.25 for each of them) and that is why the help amount they receive is almost similar for them.

**Table 4.** Received help for each type of agent and for all of the system

| Ratio | r of C | r of D | r of G | r of WG | Total r |
|---|---|---|---|---|---|
| 100C:100D:100WG:100G | 8486 | 8377 | 8388 | 8565 | 33816 |
| 25C:125D:125WG:125G | 6297 | 6353 | 6583 | 6426 | 25659 |
| 125C:25D:125WG:125G | 10624 | 10769 | 10774 | 10702 | 42872 |
| 125C:125D:25WG:125G | 6235 | 6339 | 6316 | 6264 | 25154 |
| 125C:125D:125WG:25G | 9813 | 9758 | 9983 | 9734 | 39285 |

We manipulated the ratios between agents to explore what happens with the system when there are less C, D, WG, or G agents. Most importantly, we compared the total amount of help received for the five different agent rations (Total $r$; Table 4). Confronting the help received by the system to the situation when the ratio is equal for all four type of agents, there is .27 times more help received when there are less D agents, .16 times more help received when there are less G agents, .24 less help when there are less C agents, and .26 times less help when there are less WG agents.

# 4 Discussion

In terms of a learner, these type of results should simplify a lot the understanding of the influence that each agent's behavior has on the system. It would be impossible to measure the amount of given help in a system when there are hundreds of agents and their behaviors depend on the dynamics of the whole system. For this reason, we consider ABMs one of the best learning techniques, particularly for the visualization of results related to social behaviors.

We decided to insert equal probabilities for the received help for all four type of agents, which could be considered unfair from a moral point of view (for example, D agents receive same as C agents, while they do not give help at all to others). Maybe in this situation learners can realize that whatever they decide to do about giving will not affect their receiving help, and this conclusion should puzzle them if they believe in a just world standpoint. They can start thinking about different way of behaving in a similar system in order to match their moral criteria. For example, what if the other three types of agents develop a memory variable and start not giving help anymore to D agents? This logic would not allow C agents to be pure cooperators anymore, but, at least, this should help to make people think more about the consequences of every social choice.

Considering the importance of helping behavior in a society where resources are not equally distributed and we need to exchange them in order to give a chance to everyone to survive, ABM creates an easier way to explore and understand the consequences of prosocial choices on the social system. Furthermore, the consequences of multiple agent interactions over time can be often different from the sum of the properties of each individual agent. This model could be easily implemented in a teaching context, and should simplify the learning process related to prosocial behavior in specific, but also other social situations where the people's interactions over time can provoke different results in their living environment.

The present ABM simulation is a simple example that describes helping behavior based on two basic theories, but this type of models can be adopted for several other social arguments. For example, there could be developed models to describe the consequences of decisions made by people in order to preserve the environment, or to show how tax cheating can evolve during time, or the risks of spreading gun fire in a population. All these issues, which present complex social phenomena, are difficult to be explored under a classic experimental research design. ABMs offer the opportunity to insert an infinite number of variables in a virtual society, and simulate social dynamics that can explain real behavior.

# References

1. Andreoni, J.: Impure altruism and donations to public goods: A theory of warm-glow giving. The Economic Journal 100, 464–477 (1990)
2. Axelrod, R., Hamilton, W.D.: The evolution of cooperation. Science 211(4489), 1390–1396 (1981)
3. Batson, C.D., Oleson, K.C.: Current status of the empathy-altruism hypothesis (1991)

4. Baylor, A.L.: Agent-based learning environments as a research tool for investigating teaching and learning. Journal of Educational Computing Research 26(3), 227–248 (2002)

5. Baylor, A.L., Kozbe, B.: A Personal Intelligent Mentor for Promoting Metacognition in Solving Logic Word Puzzles (1998)

6. Bonabeau, E.: Agent-based modeling: Methods and techniques for simulating human systems. Proceedings of the National Academy of Sciences of the United States of America 99(suppl. 3), 7280 (2002)

7. Brandt, H., Sigmund, K.: The logic of reprobation: assessment and action rules for indirect reciprocation. Journal of Theoretical Biology 231(4), 475–486 (2004)

8. Chen, S.-H., Yeh, C.-H.: Evolving traders and the business school with genetic programming: A new architecture of the agent-based artificial stock market. Journal of Economic Dynamics and Control 25(3-4), 363–393 (2001), doi:http://dx.doi.org/10.1016/S0165-1889

9. Cialdini, R.B., Schaller, M., Houlihan, D., Arps, K., Fultz, J., Beaman, A.L.: Empathy-based helping: Is it selflessly or selfishly motivated? Journal of Personality and Social Psychology 52, 749–758 (1987)

10. Gouldner, A.W.: The norm of reciprocity: A preliminary statement. American Sociological Review, 161–178 (1960)

11. Lotem, A., Fishman, M.A., Stone, L.: From reciprocity to unconditional altruism through signalling benefits. Proceedings of the Royal Society of London. Series B: Biological Sciences 270(1511), 199–205 (2003)

12. Macy, M.W., Willer, R.: From factors to actors: Computational sociology and agent-based modeling. Annual Review of Sociology, 143–166 (2002)

13. Malone, T.W., Lai, K.-Y., Grant, K.R.: Agents for information sharing and coordination: A history and some reflections. Paper presented at the Software Agents (1997)

14. Nowak, M.A., Sigmund, K.: The dynamics of indirect reciprocity. Journal of Theoretical Biology 194(4), 561–574 (1998)

15. Smith, K.D., Keating, J.P., Stotland, E.: Altruism reconsidered: The effect of denying feedback on a victim's status to empathic witnesses. Journal of Personality and Social Psychology 57(4), 641 (1989)

16. Trivers, R.L.: The evolution of reciprocal altruism. Quarterly Review of Biology, 35–57 (1971)

# Towards Animated Visualization of Actors and Actions in a Learning Environment

Oleksandr Kolomiyets and Marie-Francine Moens

Department of Computer Science KU Leuven,
Celestijnenlaan 200A, 3001 Leuven, Belgium
{oleksandr.kolomiytes,sien.moens}@cs.kuleuven.be

**Abstract.** This paper describes ongoing research focused on natural language understanding and visualization of actors and actions extracted from narrative text. The technique employs a natural language processing pipeline for sophisticated syntactic and semantic analysis of text, and extracts information about events, actors and their roles in events, as well as temporal ordering of the events and spatial roles. This kind of information is traditionally considered indicative for text comprehension skill tests with novel readers. The visualization is implemented in the 3D graphics prototyping environment Alice, which provides a set of visual primitives and instructions for interactions and spatial manipulations of primitives.

**Keywords:** Natural language processing and understanding, semantic analysis, text-to-scene translation, visualization.

## 1 Introduction

Ongoing developments in the digital world poses a number of challenges. One of them is the increasing degree of text complexity. The user is exposed to more and more information in textual form for which advanced assistance techniques, such as automatic illustration generation and visualization of the textual content, might help the users to better comprehend it.

This paper describes ongoing research in visualization of simple narrative texts. The main contribution of the paper is a deep semantic analysis of text provided by the natural language processing pipeline, which involves recent developments in temporal and spatial information extraction. These semantic annotations are used for visualizing the recognized actions and the corresponding actors generated in a 3D graphics prototyping environment, and the actors and actions are rendered as animations. The results of this work that comprises manual and semi-automatic story animations are made available.

## 2 Related Work

In the last decade there has been a substantial amount of work in the domain of text-to-scene translation. The WordsEye project [1] focuses on an automated

T. Di Mascio et al. (eds.), *Methodologies and Intelligent Systems for Technology Enhanced Learning*, Advances in Intelligent Systems and Computing 292,
DOI: 10.1007/978-3-319-07698-0_25, © Springer International Publishing Switzerland 2014

approach to generation of scenes from natural language. Johansson et al. [2] describe an automatic text-to-scene conversion system for the traffic accident domain for Swedish. A similar work for French is also known [3]. Joshi et al. [4] propose an automatic text illustration system which automatically extracts keywords from text and relevant pictures are found in a database. Mihalcea and Leong [5] present a system for the automatic construction of pictorial representations for simple sentences. Zhu et al. [6] propose a Text-to-Picture system that generate a picture from a natural language text. Bui et al. [7] develop an application for automated translation of natural language patient instructions into pictures.

Our work is different to all the previous works in many aspects. First, we target the visualization task by producing animations and not static pictorial illustrations. Second, in contrast to the previous work, we do not rely on some sort of controlled-language descriptions, we use natural language texts. And third, we employ a number of wide-range semantic annotations extracted from text, including temporal and spatial annotations.

# 3   Natural Language Processing Pipeline

The visualization procedure that receives an input narrative and provides a visual representation of actions and involved actors relies on a natural language processing pipeline. This pipeline makes use of a number of automated syntactic and semantic analyses produced by external processing tools (for syntactic processing) as well as domain-specific algorithms and recognition models. Let us consider the following narrative:

> *Two Travellers were on the road together, when a Bear suddenly appeared on the scene. Before he observed them, one made for a tree at the side of the road, and climbed up into the branches and hid there. The other was not so nimble as his companion; and, as he could not escape, he threw himself on the ground and pretended to be dead. The Bear came up and sniffed all round him, but he kept perfectly still and held his breath: for they say that a bear will not touch a dead body. The Bear took him for a corpse, and went away. When the coast was clear, the Traveller in the tree came down, and asked the other what it was the Bear had whispered to him when he put his mouth to his ear. The other replied, "He told me never again to travel with a friend who deserts you at the first sign of danger."*

**Fig. 1.** A narrative story from the Aesop's fable corpus

## 3.1   Event Recognition

In the NLP domain there are a number of definitions that regard events. In the Automatic Content Extraction context, events were defined as a complex structure with arguments. Each ACE event relates a predefined searchable topic of interest with arguments and a set of argument roles. A more honed definition of events was presented by Filatova and Hatzivassiloglou [8] which addressed some semantic representation for interpretations of "Who did what to whom when and where?" and was adopted in the novel annotation standard for temporal events, temporal expressions and temporal relations, and proposed in the TimeML markup annotation language [9]. In this work we also follow the

TimeML-based definition of events as the most relevant representation of actions and their visualizations[1]. Recent annotation efforts have shown that a simplified set of TimeML-based annotation guidelines has been found more practical and resulted in a higher interannotator agreement [10]. Moreover, such an annotation approach has given rise to sophisticated structured-output algorithms for temporal information processing [11].

## 3.2    Entity Recognition

Annotation of people, locations and other named entities for natural language processing is typically traced back to the Message Understanding Conferences (MUC), where the named entity task included identifying proper nouns in a text that referred to organizations, persons and locations [12]. In this work we primarily focus on the recognition of persons and locations as the central entities for action visualizations.

## 3.3    Event Participants Roles Recognition

Links from entities to the events they participate in have been annotated in a variety of different forms ([13–15]). In the scope of the described research with the goal of visualization of actions and actors, we primarily focus on action actors that are identified at the positions of syntactic subjects and objects for verb-triggered events. This information can be provided by dependency parsers [16] or derived from the syntactic constituent parse [17].

## 3.4    Coreference Resolution

Two words or phrases in a text are said to co-refer if they are both references the same object or person mentioned in the discourse. Discourse objects are mapped by the reader into elements of a mental model, that may or may not correspond to actual entities or events of the real world. In the full narrative discourse, co-referential links would form coreference chains where all mentions of a chain refer to the same entity. In computational linguistics a number of approaches to coreference resolution have been proposed. One of the most successful approaches with a high performance level is described in [18].

## 3.5    Temporal Ordering of Events

Narratives are a kind of texts that have a story plot and exhibit a particular structure of temporally ordered events, that is, a connected graph with nodes as events and edges as relations between the events labeled with tags such as *before* or *after* with respect to the temporal order in which the events occur. With the advent of TimeML, three major concepts of temporal information in text were

---

[1] In this work we refer to actions as to a subset of temporally-anchored events as physical involvements of agents (protagonists) and interactions between them.

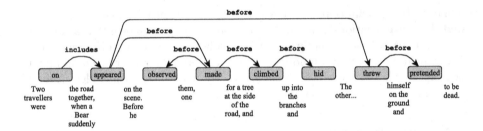

**Fig. 2.** A timeline for the narrative in Fig. 1. Nodes are events and edges are temporal relations signaled by linguistic cues in the text.

defined: events, timexes (times) and temporal relations. Being thoroughly studied in the domain of newswire processing (e.g., [19]), temporal annotations of narrative timelines have been recently investigated in [11], which can derive temporal structures for narratives in terms of dependency relations that constitute a timeline, as presented in Fig. 2. In our work we employ this method.

### 3.6 From Events to Actions

So far we have described the tasks of semantic information extraction and natural language processing that are required for an automated "machine-based" language understanding system. For visualization purposes however, not all recognized events are relevant or "visualizable". Let us examine semantic classes of events. With respect to semantic classification of events, TimeML defines seven categories: reporting, perception, aspectual, intensional actions and states, states and occurrences. As Bethard et al. pointed out [10], human-based annotation of events and their semantic classes is a very difficult task even for experts in TimeML, and thus, they introduced the notion of *factual events* and *states*. Let us examine the events from the example presented in Fig. 1. One can distinguish the following (non-disjunctive) groups of events:

- factual events: *were **on** the road ..., suddenly **appeared** on the scene ...*, etc.
- events in possible worlds: *he could not **escape**..., a bear will not **touch** ...*
- states: *was not so **nimble** as ..., the coast was **clear**, ...*
- motions: *suddenly **appeared** on the scene ..., the bear **came up** ...*

As the first three groups are the instances of the TimeML semantic event classification scheme, motions are a quite novel class of events that recently has triggered a specific research interest. The most recent advances in this field are the SpatioTemporal Markup Language (STML) [20], and the Motion Markup Language (MotionML) [21]. With the similar objective, STML and MotionML were designed for capturing motions in text. While STML focuses on annotating motions with fine-grained semantics of spatial entities and their roles for further qualitative spatial reasoning, MotionML proposes a shallow annotation approach, where *trajector*'s motions are annotated along with a single spatial label *path* triggered by *motion indicators*. The simplified motion annotation approach comes at the expense of the spatial semantic granularity, however, such

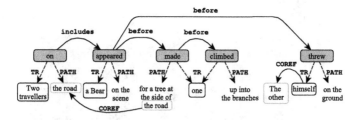

**Fig. 3.** A story timeline for the narrative in Fig. 1 with events considered for actions. Every action represents a motion in the story with argument roles such as *trajector* (TR) and *path* (PATH). Coreferential links are labeled with (COREF).

an approach does not restrict the expressiveness of the annotation language and its potential scalability to STML. MotionML was proposed as an extension for the set of spatial roles introduced by the Spatial Role Labeling schema, and successfully adapted to the Spatial Role Labeling task [22]. For visualization purposes, we only focus on motion indicators and use them as a dictionary of visual actions. Fig. 3 represents a spatio-temporal structure of the narrative.

## 4  The Prototyping Environment Alice

Alice[2] is a 3D graphics prototyping environment developed for teaching and extending computational thinking for students of different age groups and backgrounds. It provides a number of animated and non-animated graphical primitives (persons, animals, characters, vehicles, scenery objects, etc.), which can be placed in a virtual world with customized properties such as color, position opacity, and size. Let us summarize a list of procedural methods applicable to an object in Alice: (i) conversational procedures (say and think); (ii) orientations (turn, roll, turnToFace, etc.); (iii) positions (move, moveToward, moveAwayFrom, etc.); (iv) appearance (setOpacity and setPaint). Yet, Alice provides a number of object functional methods, i.e. the methods which return the object's property value (similar to Java getters), for example: getWidth(), getPaint(), but also spatial properties, such as isFacing(obj), getDistanceTo(obj).

Since animated objects are internally represented as one of the skeletal joint systems, orientation procedures can also be applied to the parts of the skeletal joint system to model. Yet, Alice provides a set of programming statements such as do_in_order, do_together, while, for_each and allows one to create visualization/interaction scripts for visualizing complex events and interactions composed by atomic Alice instructions with respect to the user-defined logic.

## 5  Mapping Annotations to Graphical Primitives

After actions and actors in the text have been recognized by the NLP pipeline, we use these annotations to populate the virtual world in Alice. A number of

---

[2] http://www.alice.org

**Fig. 4.** The initial story setting with three actors (a Cat and a Hare as Travelers, and a Panda as the Bear). The palm tree represents *a tree at a side of the road.*

assumptions, however, are made: (i) the provided annotations are disambiguated in terms or their meaning, i.e., they are mapped to a single synset in the lexical resource WordNet [23]; (ii) we manually determine visual appearance of characters in Alice; (iii) the spatial setting of the scene is predetermined. The initial spatial layout of the scene is presented in Fig. 4. After the spatial scene has been initialized, we generate Alice procedures from the annotations of actors and actions. The procedure is based on the mapping actions (and their actors) to a set of motion and conversational procedures. The lexical diversity of actions in text is treated by defining the root concepts in WordNet (`give voice`, `formulate#3`, and `state`, `say#1` for *saying*, and `travel`, `go#1` for *moving*), so that if an action annotation cannot be directly matched with either a move or a say action, a search algorithm checks if the lexical item can be found in the list of synonyms for root actions or the root action is an inherit hypernym for the action in question. Each action receives the actor and a number of additional parameters, such as strings for conversational procedures, or absolute/relative directions and distances for motions. Let us look at the instruction for the actions of a traveler (the cat) moving to the tree and climbing it:

```
\\the cat is moving to the tree
this.cheshireCat.moveToward(
  this.coconutPalm, \\motion path, target
  this.cheshireCat.getDistanceTo(this.coconutPalm)); \\distance
\\the cat is climbing the tree
this.cheshireCat.move(
  MoveDirection.UP, \\motion direction
  this.coconutPalm.getHeight()*0.75); \\motion distance
```

where `this.cheshireCat` is the (actor) object and `this.coconutPalm` is the path (target) of the motion.

Once visual instructions in Alice have been generated, temporal relations between events provided by the NLP pipeline specify the temporal order for execution of the instructions. As the recognized temporal relation labels are merely text strings, they have to be transformed into some computational form. One of the computational forms for representing temporal relations is the Allen's temporal interval algebra. We first select actions which are in `includes` or `is_included` relations to each other and generate a `do_together` instruction block that includes them. Temporal relations with the label `identity` signal merging temporal events to their antecedents. After that, other temporal relations with labels `before` and `after` are inverted to a single label `after` for generating a linear order of instruction execution in Alice. Two animations are available[3]: (i) manually created, and (ii) generated semi-automatically.

# 6  Discussion and Conclusions

In this paper we have presented the first prototype for narrative visualizations. The technique employs a natural language processing pipeline for syntactic and semantic analysis of text. At the semantic level it provides information about actions and actors, their relations, temporal ordering of the actions on the timeline, and spatial roles for the identified actions. The visualization technique translates semantic annotations into instructions in the visualization environment by mapping these annotations into visual primitives with a number of assumptions. We recognize these assumptions as bottlenecks for a generic and fully domain-independent visualization procedure, and would like to provide the reader with the emerging directions that address these challenges.

**Temporal ordering** is essential for text comprehension and received great interest in the research community. In language, temporal ordering is implemented by aspectual and tensed cues. However, in many cases this information is not enough to provide an accurate temporal analysis since most of the cues reside in commonsense knowledge. *Event durations* are very important information for visualization, however, there is very little evidence in text about them. This information humans receive from other sources (e.g. personal experience and observations). The lack of duration information prevents us from exact visual replication of the story plot (and also using TimeGraph [24] as the computational means for timelines), however, for the chosen text genre and text complexity, the visualized story reached a good approximation level of the plot semantics.

**Spatial information** is a type of semantic annotations vital for visualizations. All visualizations techniques are designed around motions. At the same time spatial information is always present in narratives, but an automated spatial analysis of text remains a far reaching goal. Two important research initiatives addressed spatial annotations in text: STML and MotionML. In this work we used MotionML and the corresponding annotated corpus for identifying motion actions in text. STML, on the other hand, is designed for annotating fine-grained spatial semantics, and seems to be more powerful for consequent qualitative spatial reasoning, but, at this moment, no annotations in STML are available. Spatial annotations are very important for visualizing the initial spatial scene, but

---

[3] http://people.cs.kuleuven.be/~oleksandr.kolomiyets/ebuTEL2013.zip

similarly to temporal ordering, this information is usually not directly available in text, where personal commonsense knowledge primes this visualization.

**Acknowledgments.** The presented research was supported by the TERENCE (EU FP7-257410) and MUSE (EU FP7-296703) projects.

# References

1. Coyne, B., Sproat, R.: WordsEye: an automatic text-to-scene conversion system. In: Proceedings of the 28th Annual Conference on Computer Graphics and Interactive Techniques, pp. 487–496. ACM (2001)
2. Johansson, R., Berglund, A., Danielsson, M., Nugues, P.: Automatic text-to-scene conversion in the traffic accident domain. In: Proceedings of the International Joint Conference on Artificial Intelligence, pp. 1073–1078 (2005)
3. Kayser, D., Nouioua, F.: From the textual description of an accident to its causes. Artificial Intelligence 173(12), 1154–1193 (2009)
4. Joshi, D., Wang, J.Z., Li, J.: The story picturing engine—a system for automatic text illustration. ACM Transactions on Multimedia Computing, Communications, and Applications 2(1), 68–89 (2006)
5. Mihalcea, R., Leong, C.W.: Toward communicating simple sentences using pictorial representations. Machine Translation 22(3), 153–173 (2008)
6. Zhu, X., Goldberg, A.B., Eldawy, M., Dyer, C.R., Strock, B.: A text-to-picture synthesis system for augmenting communication. In: Association for the Advancement of Artificial Intelligence, pp. 1590–1595 (2007)
7. Bui, D., Nakamura, C., Bray, B.E., Zeng-Treitler, Q.: Automated illustration of patients instructions. In: Proceedings of the Annual Symposium of the American Medical Informatics Association, AMIA, pp. 1158–1167 (2012)
8. Filatova, E., Hatzivassiloglou, V.: Domain-independent detection, extraction, and labeling of atomic events. In: Proceedings of the Conference on Recent Advances in Natural Language Processing (2003)
9. Pustejovsky, J., Castaño, J., Ingria, R., Saurí, R., Gaizauskas, R., Setzer, A., Katz, G.: TimeML: Robust specification of event and temporal expressions in text. In: Proceedings of the International Workshop on Computational Semantics (2003)
10. Bethard, S., Kolomiyets, O., Moens, M.F.: Annotating narrative timelines as temporal dependency structures. In: Proceedings of the International Conference on Linguistic Resources and Evaluation (2012)
11. Kolomiyets, O., Bethard, S., Moens, M.F.: Extracting narrative timelines as temporal dependency structures. In: Proceedings of the 50th Annual Meeting of the Association for Computational Linguistics, pp. 88–97. ACL (2012)
12. Chinchor, N., Robinson, P.: MUC-7 Named entity task definition. In: Proceedings of the 7th Conference on Message Understanding (1997)
13. Marcus, M.P., Marcinkiewicz, M.A., Santorini, B.: Building a large annotated corpus of English: The Penn Treebank. Computational linguistics 19(2) (1993)
14. Baker, C.F., Fillmore, C.J., Lowe, J.B.: The Berkeley FrameNet project. In: Proceedings of the 17th International Conference on Computational Linguistics, pp. 86–90. ACL (1998)
15. Palmer, M., Gildea, D., Kingsbury, P.: The proposition bank: An annotated corpus of semantic roles. Computational Linguistics 31(1), 71–106 (2005)

16. Nivre, J., Hall, J., Nilsson, J.: Maltparser: A data-driven parser-generator for dependency parsing. In: Proceedings of LREC, pp. 2216–2219 (2006)
17. De Marneffe, M.C., Manning, C.D.: The Stanford typed dependencies representation. In: Proceedings of the Workshop on Cross-Framework and Cross-Domain Parser Evaluation, pp. 1–8. ACL (2008)
18. Lee, H., Peirsman, Y., Chang, A., Chambers, N., Surdeanu, M., Jurafsky, D.: Stanford's multi-pass sieve coreference resolution system at the CoNLL-2011 shared task. In: Proceedings of the Fifteenth Conference on Computational Natural Language Learning: Shared Task, pp. 28–34. ACL (2011)
19. Verhagen, M., Saurí, R., Caselli, T., Pustejovsky, J.: SemEval-2010 Task 13: TempEval-2. In: Proceedings of the 5th International Workshop on Semantic Evaluation, pp. 57–62 (2010)
20. Pustejovsky, J., Moszkowicz, J.L.: The qualitative spatial dynamics of motion in language. Spatial Cognition & Computation 11(1), 15–44 (2011)
21. Kolomiyets, O., Moens, M.F.: MotionML: A shallow approach for annotating motions in text. In: Proceedings of Corpus Linguistics (2013)
22. Kolomiyets, O., Kordjamshidi, P., Moens, M.F., Bethard, S.: SemEval-2013 Task 3: Spatial Role Labeling. In: Proceedings of the Seventh International Workshop on Semantic Evaluations, pp. 255–262. ACL (2013)
23. Miller, G.A.: WordNet: a lexical database for English. Communications of the ACM 38, 39–41 (1995)
24. Miller, S.A., Schubert, L.K.: Time revisited. Computational Intelligence 6(2), 108–118 (1990)

# Author Index